Basic Principles of Nanotechnology

Basic Principles of Nanotechnology

Wesley C. Sanders

CRC Press is an imprint of the
Taylor & Francis Group, an **informa** business

CRC Press
Taylor & Francis Group
6000 Broken Sound Parkway NW, Suite 300
Boca Raton, FL 33487-2742

© 2019 by Taylor & Francis Group, LLC
CRC Press is an imprint of Taylor & Francis Group, an Informa business

No claim to original U.S. Government works

Printed on acid-free paper

International Standard Book Number-13: 978-1-138-48361-3 (Hardback)
978-1-138-48828-1 (Paperback)

This book contains information obtained from authentic and highly regarded sources. Reasonable efforts have been made to publish reliable data and information, but the author and publisher cannot assume responsibility for the validity of all materials or the consequences of their use. The authors and publishers have attempted to trace the copyright holders of all material reproduced in this publication and apologize to copyright holders if permission to publish in this form has not been obtained. If any copyright material has not been acknowledged please write and let us know so we may rectify in any future reprint.

Except as permitted under U.S. Copyright Law, no part of this book may be reprinted, reproduced, transmitted, or utilized in any form by any electronic, mechanical, or other means, now known or hereafter invented, including photocopying, microfilming, and recording, or in any information storage or retrieval system, without written permission from the publishers.

For permission to photocopy or use material electronically from this work, please access www.copyright.com (http://www.copyright.com/) or contact the Copyright Clearance Center, Inc. (CCC), 222 Rosewood Drive, Danvers, MA 01923, 978-750-8400. CCC is a not-for-profit organization that provides licenses and registration for a variety of users. For organizations that have been granted a photocopy license by the CCC, a separate system of payment has been arranged.

Trademark Notice: Product or corporate names may be trademarks or registered trademarks, and are used only for identification and explanation without intent to infringe.

Visit the Taylor & Francis Web site at
http://www.taylorandfrancis.com

and the CRC Press Web site at
http://www.crcpress.com

For Wesli

Contents

Preface

Nanotechnology is an integrated discipline that includes concepts from chemistry, physics, materials science, and engineering. Nanotechnology specifically focuses on matter 1–100 nanometers in size. For reference, a single human hair is about 100,000 nanometers across. While speaking to Congress in 1999, Nobel prize laureate Richard Smalley expressed the importance of this field by stating, "the impact of nanotechnology on the health, wealth, and lives of people will be at least the equivalent of the combined influences of microelectronics, medical imaging, computer-aided engineering, and man-made polymers" (Rogers, Adams and Pennathur 2013). The National Science Foundation estimates nanotechnology has a three trillion-dollar impact on the global economy, and in addition requires the employment of six million workers (Morris et al. 2015). For this reason, several two and four-year colleges now offer nanotechnology coursework and training programs. The purpose of this book is to provide a general description of basic nanomaterials, characterization tools, and fabrication processes routinely used in academia and industry. This book serves as a reference for students planning to pursue coursework in nanotechnology. Students preparing for roles as technicians or entry level scientists in nanotechnology-based laboratories will find this book useful as well. Familiarity with basic concepts from the physical sciences helps with understanding the unique behaviors occuring at the nanoscale. For this reason, two chapters covering relevant concepts from freshman-level chemistry and physics are included. Additionally, there are a few algebra-based mathematical relationships included in this book. The reader will find this book focuses primarily on key nanomaterials routinely used in academic and industrial pursuits including carbon nanotubes, graphene, metal nanoparticles, quantum dots, and polymers just to name a few. Additionally, this book describes basic characterization and fabrication techniques commonly used in nanotechnology research including scanning electron microscopy, atomic force microscopy, etch, e-beam lithography, and photolithography. To add, this book addresses the impact of nanotechnology on two major industries, the medical industry and the electronics industry. A discussion of the effects of nanotechnology on major industries is important because there are billions of nanoscale transistors on silicon wafers approximately one square inch in size. Also, the medical field utilizes nanotechnology to diagnose and treat cancer by employing the use of quantum dots, metal nanoparticles, and dendrimers.

The following concepts are addressed in this book:

- Examples of ancient artifacts incorporating nanotechnology
- Important figures in nanotechnology

- Important chemistry and physics principles associated with nanotechnology
- Unique optical, electrical, and mechanical properties of commonly-used nanomaterials
- Processes used to create, manipulate, and characterize nanomaterials
- Applications of nanotechnology in electronics and medicine

Each chapter begins with a list of key objectives. Chapters also contain several figures, tables, and illustrations to provide additional information and interest. At the end of each chapter, the reader will find several end-of-chapter questions to reinforce chapter content. Reference pages at the end of each chapter are included to provide opportunities for further reading.

References

Morris, J., P. Moeck, L. Weasel, and J. Straton. 2015. "Nanotechnology courses for general education." *Proceedings of the 122nd ASEE Annual Conference and Exposition*. Seattle.

Rogers, B., J. Adams, and S. Pennathur. 2013. *Nanotechnology The Whole Story*. Boca Raton: CRC Press/Taylor & Francis Group.

Wesley C. Sanders

Author

Wesley C. Sanders is currently an assistant professor at Salt Lake Community College. He teaches courses in nanotechnology, materials science, chemistry, and microscopy. While serving as an assistant professor, he has published articles in the *Journal of Chemical Education* describing undergraduate labs for use in introductory nanotechnology courses. He earned a BSEd in science education from Western Carolina University (1999). Later, he earned an MS in chemistry from the University of North Carolina at Charlotte (2003) and a PhD in chemistry from Virginia Tech (2008). His initial experiences with nanotechnology occurred while studying self-assembled monolayers on gold with a scanning electrochemical microscope as a doctoral student at Virginia Tech. After earning his PhD, he examined bacterial nanofilaments with an atomic force microscope while working as a postdoctoral researcher at the US Naval Research Lab in Washington DC.

1

Introduction to Nanotechnology

Key Objectives

- Know the role of nanotechnology in consumer products
- Become familiar with ancient uses of nanotechnology
- Become familiar with past researchers involved with nanotechnology
- Learn how the natural world uses nanotechnology

1.1 Introduction

The word nanotechnology was first introduced in 1974 by Norio Taniguchi. He defined nanotechnology as "processing of separation, consolidation, and deformation of materials by one atom or one molecule" (Mulvaney 2015). Furthermore, in a paper entitled "On the basic concept of Nano-Technology" he stated, "In the processing of materials, the smallest bit size of stock removal, accretion, or flow of materials is probably of one atom or one molecule, namely 0.1–0.2 nm in length" (Rogers, Adams and Pennathur 2013). Since then, scientists and engineers have defined nanotechnology as the science of matter 1 billionth of a meter (10^{-9} m) in size. This technology involves the manipulation of matter at the atomic and molecular scales (Horikoshi and Serpone 2013). Nanotechnology utilizes concepts from physics, chemistry, and materials science in efforts to explain the unique behaviors of nanoscale materials (Rogers, Adams and Pennathur 2013).

1.2 Consumer Nanotechnology

Nanoscale materials referred to as engineered nanoparticles (ENPs) are currently utilized in a wide range of commercial products (Zhang et al. 2015).

ENPs are included in products such as cosmetics, personal care products, foods (processing and packaging), clothing, and detergents just to name a few (Zhang et al. 2015). Additional examples include the use of carbon nanotubes—hollow and cylindrical tubes of carbon—for use in polymer composites, electromagnetic shielding, electron field emitters (flat panel displays), super capacitors, batteries, hydrogen storage, and structural composites (Aitken et al. 2006). Conducting or semiconductor nanowires, with diameters a few tens of nanometers in size, are used as interconnectors in nanoelectronic devices (Aitken et al. 2006). Significant advances have been achieved in the semiconductor industry due to the incorporation of nanotechnology. For instance, cell phones have transformed into devices with multiple applications (Kaiser and Kuerz 2008). With current cell phones, users can send text messages, take and send pictures, run internet applications, listen to music, play games, and watch movies (Kaiser and Kuerz 2008). This is possible because cell phones contain increasingly powerful computer chips with a substantial amount of nanoscale transistors influencing the functionality of cell phones (Kaiser and Kuerz 2008). Computers have become compact and increasingly powerful due to nanotechnology. The genesis of computer chips starts with Jack Kilby of Texas Instruments and Robert Noyce of Fairchild Semiconductor. They developed the first integrated circuit (IC) in the late 1950s. In 1961, Fairchild Semiconductor used a photoetching process to produce vast numbers of transistors on a thin slice of silicon. Component sizes on the first ICs were 5 μm in size (Madou 2011). In the early 1970s, Intel microprocessors contained 2300 transistors (El-Aawar 2015). Decades later, in 2003, Intel introduced the Pentium IV chip, with 90 nm component sizes. ICs with 65 nm component sizes became available in 2006, and ICs with 32 nm component sizes were developed in 2009 (Madou 2011). In 2014, Intel microprocessors contained 5.56 billion transistors, with 22 nm component sizes (Madou 2011).

1.3 Ancient Nanotechnology

1.3.1 Lycurgus Cup

Nanotechnology has been used in sculptures, paintings, and other artifacts since the fourth century AD. For example, the Lycurgus cup is an artifact containing dichroic glass, which is a material that changes color depending on the nature of light exposure (Horikoshi and Serpone 2013). When light is reflected off the surface of the cup, the cup appears green, however, when light passes through the cup it appears red (Freestone et al. 2007). This unusual optical effect is a direct result of the presence of a small quantity of 70 nm diameter particles of silver and gold in the glass (Horikoshi and Serpone 2013). The gold component is responsible for the red color and the silver component is responsible for the green color (Freestone et al. 2007).

1.3.2 Damascus Swords

Damascus swords are ancient artifacts that, according to reports, can cut a piece of silk in half as it falls to the ground. These swords possess high mechanical strength, flexibility, and sharpness. It is claimed that the blades can be bent at angles up to 90°. To explain the high mechanical properties and flexibility of the swords, a specimen was taken from one of the swords and dissolved in hydrochloric acid. The remnants were examined by high-resolution transmission electron microscopy, and the images revealed the presence of carbon nanotubes that were not dissolved by the acid. Artisans created the swords utilizing heating and forging steps that incorporated organic debris. The debris assisted in the formation of the carbon nanotubes found in the swords (Srinivasan 2007).

1.3.3 Stained Glass Windows

There is also evidence of nanotechnology in the Middle Ages, particularly in the stained-glass windows used in churches during that time. During the Middle Ages, it was common to use a ruby red color in stained glass windows. Beautiful examples of this application can be seen in the windows of the Cathedral Notre-Dame de Chartres in France (Figure 1.1) (Horikoshi and Serpone 2013). Medieval artisans were unaware of the scientific explanation regarding the colors of the glass; they were simply aware of the color produced by mixing gold chloride into molten glass. This resulted in the creation of tiny gold spheres, which absorbed and reflected sunlight in a way that produced a ruby-red color. The gold nanoparticles were 25 nm in diameter (Chang 2005). When the nanoparticles in stained glass interact with light, certain wavelengths of light are absorbed via a surface plasmon resonance mechanism (a topic covered in Chapter 6), while other wavelengths of light are reflected. This effect was unknown during the Middle Ages. However, the medieval artisans knew that introducing different metals to the glass altered the color of the glass. They knew gold produced a deep ruby red, copper produced blue or green, and iron produced green or brown (Kolwas, Derkachova and Jakubczyk 2016).

1.4 Early Nanotechnologists

1.4.1 Michael Faraday

Previous sections describe ancient, practical uses of nanotechnology. It wasn't until the 1800s that scientists, such as Michael Faraday, began to search for the impetus behind the unique properties of nanomaterials. Michael Faraday experimented with a variety of metal nanoparticles suspended in water

FIGURE 1.1
Notre Dame Cathedral North Rose window. The Rev. Paul Cioffi, S.J. Photographs Collection, Georgetown University.

(Thompson 2007). In 1857, Michael Faraday reported the production of ruby red mixtures containing gold nanoparticles; he referred to these mixtures as colloidal gold (Horikoshi and Serpone 2013). A colloidal mixture, or colloid, is a mixture of two phases of matter. These mixtures are usually solid particles dispersed in a liquid or gas. The suspended particles can absorb or scatter visible light (Rogers, Adams and Pennathur 2013). When Faraday added a reducing material to sodium tetrachloroaurate ($NaAuCl_4$), he observed that the yellow color of $NaAuCl_4$ changed over a course of a few minutes to a deep ruby red color. Faraday suggested the ruby-red mixture contained very fine metallic gold not visible with any of the microscopes available during his time (Thompson 2007). Nearly 100 years later, electron microscopy was used to reveal that ruby-red colloids contain particles of gold with diameters approximately 6 nm in size (Thompson 2007).

1.4.2 Richard Feynman

Richard Feynman was a Nobel prize-winning physicist who presented a now famous lecture on atom-by-atom assembly. This lecture is often credited with kick-starting nanotechnology (Ball 2009). The lecture Feynman presented to the American Physical Society on December 29th, 1959 was entitled "There's plenty of room at the bottom." Among the things he discussed, was the possibility of small devices. He states, *"I don't know how to do this on a small scale in a practical way, but I do know that computing machines are very large; they fill rooms. Why can't we make them very small, make them of little wires, little elements – and by little I mean little. For instance, the wires should be 10 or 100 atoms in diameter, and the circuits should be a few thousand angstroms across. There is plenty of room to make them smaller. There is nothing that I can see in the laws of physics that says the computer elements cannot be made enormously smaller than they are now."* (Roukes 2001). Additionally, in this speech Feynman discussed the amount of space required to store written material on the nanoscale, *"For each bit I allow 100 atoms. And it turns out that all of the information that man has carefully accumulated in all the books in the world can be written in this form in a cube of material one two-hundredth of an inch wide – which is the barest piece of dust that can be made out by the human eye. So there is plenty of room at the bottom! Don't tell me about microfilm! (Baird and Shew 2004).* Feynman's vision was realized 30 years later (Ball 2009).

1.4.3 K. Eric Drexler

A few decades after Richard Feynman's speech regarding atomic-level manipulation, K. Eric Drexler published a book entitled, *Engines of Creation* (1986). In this book, Drexler described "molecular technology," which involves creating structures using atoms and molecules (Pradeep 2007). Drexler believed molecular nanotechnology provides engineers with the ability to create nanocircuits and nanomachines (Drexler 1986). Additionally, he believed molecular assembly would be driven by ordinary chemical reactions (guided by other nanomachines). In his book, Drexler also predicted biochemists would use protein molecules as motors, bearings, and moving parts to build robot arms capable of handling individual molecules (Drexler 1986). Drexler also suggested that bonded atoms resemble bearings mounted by single chemical bonds allowing atoms to turn freely and smoothly (Drexler 1986).

1.4.4 Donald M. Eigler (IBM)

The predictions of Richard Feynman and K. Eric Drexler regarding atomic and molecular scale manipulation were realized in 1989. In that year, Dr. Donald M. Eigler of IBM's Almaden Research Center in San Jose, California, positioned individual xenon atoms spelling out "IBM" (Figure 1.2)

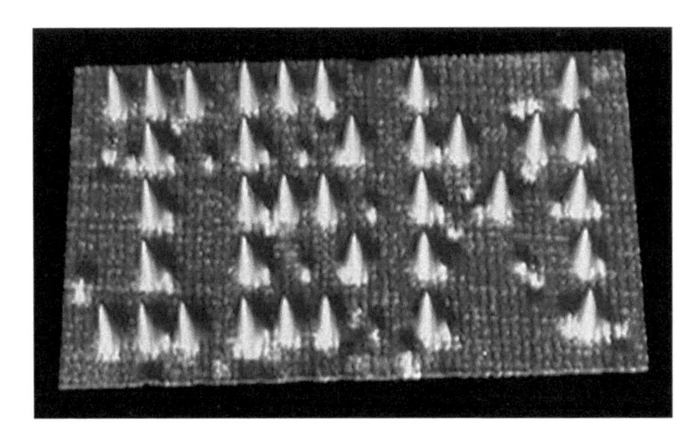

FIGURE 1.2
IBM company logo spelled out with 35 xenon atoms on nickel. Image originally created by IBM Corporation.

(Browne 1990). This was accomplished by depositing xenon atoms on the surface of a cold nickel surface. Liquid helium was used to chill the sample 452 degrees below zero Fahrenheit to reduce the motion of the xenon atoms. This facilitated easier manipulation of the randomly arranged xenon atoms with a scanning tunneling microscope (Browne 1990). When the scanning tunneling microscope tip was a certain distance above a xenon atom, the needle attracted the atom. This allowed Eigler to drag the xenon atom across the surface of the nickel and place it in any desired position. The xenon atom was released by slightly raising the scanning tunneling microscope tip (Browne 1990).

1.5 Nanotechnology in Nature

1.5.1 Insect Colors

The use of nanotechnology in art, weaponry, and scientific research was discussed in the previous sections. However, it is interesting to point out that the natural world uses nanotechnology. For instance, bright and vivid colors seen in nature oftentimes arise from the interaction of light with periodically arranged, micro- and nanoscale structures. The wings of morpho butterflies contain a combination of multilayer optical gratings and other unique structures (Figure 1.3), which produce an array of complex colors. It is interesting to point out that these structures are wavelength-selective and can reflect specific wavelengths over a broad range of angles (Kolle et al. 2010). Researchers have discovered that certain species of tarantulas, such as *P. metallica* and *L. violaceopes* (blue tarantulas), contain multilayer

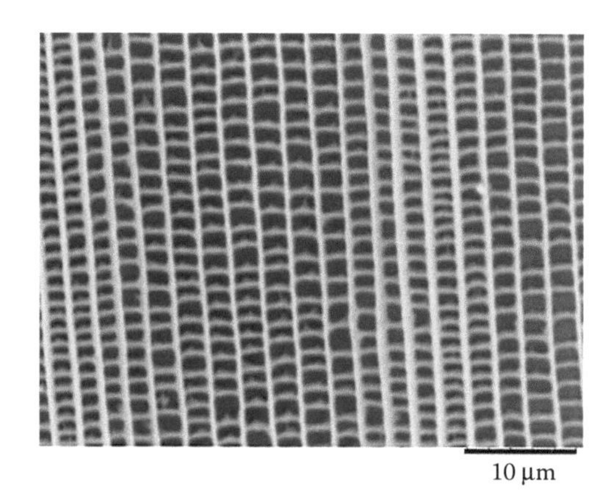

FIGURE 1.3
Scanning electron microscope image of a butterfly wing.

nanostructures marked with cylindrical groove-like structures. The presence of these structures produces a blue iridescence on the body of the insect (Hsiung, Blackledge and Shawkey 2014).

1.5.2 Geckos

Geckos can climb and run on wet, dry, smooth, or rough surfaces with extremely high maneuverability and efficiency. Gecko feet (Figure 1.4a) contain compliant micro- and nanoscale beta-keratin structures, known as foot-hairs, which allow geckos to adhere to any surface. This adhesion is driven by molecular forces acting between the foot-hairs and the surface (Autumn et al. 2002; Autumn, Niewiarowski and Puthoff 2014). Foot-hairs contain micrometer-scale stalks with caps at their end—referred to as spatulae, which have diameters that are approximately 300–500 nm in diameter (Figure 1.4b) (Sitti and Fearing 2003).

FIGURE 1.4
Gecko foot (a) and microscale foot-hairs (b). Images originally created by Dr. Autumn Kellar.

1.5.3 Hydrophobic Surfaces

Hydrophobic surfaces cause water to readily bead up and roll off without wetting them (Verbanic et al. 2014). There are several examples of surfaces in nature exhibiting hydrophobicity due to the presence of nanoscale structures. The lotus plant growing at the bottom of ponds emerges above the water surface while remaining untouched by the contaminants in the dirty water. Water droplets roll over the leaf's surface taking away all the dirt and leaving a clean surface behind; this is referred to as the lotus effect (Balani et al. 2009). The lotus effect is generated by an organized arrangement of microscale and nanoscale spires covering the surface of the leaf (Figure 1.5) (Verbanic et al. 2014). A similar hydrophobic effect is observed with cicadas due to the presence of arrays of nanosized pillars 100 nm in diameter and 300 nm in height on the wings of the insects (Hong, Hwang and Lee 2009).

1.5.4 Photonic Phenomena

The Attacus atlas moth uses a complex imaging system with compound lenses used to accommodate low brain processing capabilities. The compound lens system contains hexagonally shaped ommatidia, each containing its own optical microlens. Similar structures are found in the compound eyes of houseflies (Figure 1.6). This allows the production of individual images, which provide insects with a large field of view without the need of increased eye volume. Interestingly, each ommatidium possesses unique nanometer-scale surface structures behaving as antireflective surfaces. This enables insect eyes to have photosensitivity in dim environments in addition to reducing reflections otherwise visible to predators (Ko et al. 2011).

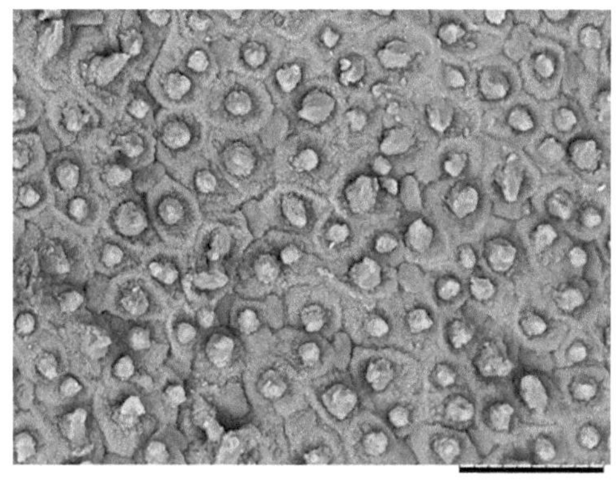

FIGURE 1.5
Scanning electron microscope image of microscale protrusions on the surface of a lotus leaf.

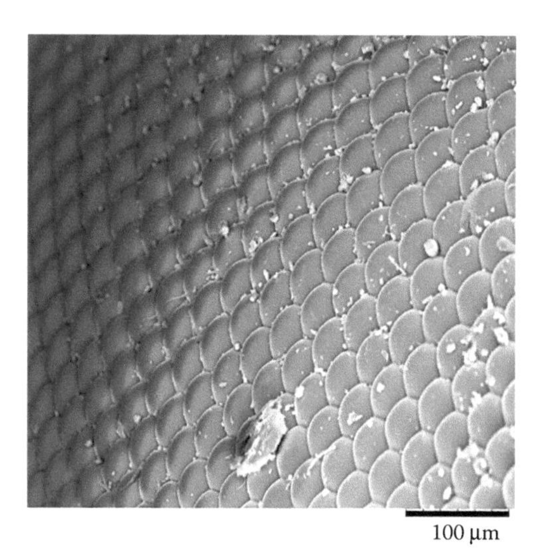

FIGURE 1.6
Scanning electron microscope image of a fly eye.

1.6 End-of-Chapter Questions

1. In 1974, Norio Taniguchi stated nanotechnology involves the processing of materials by:
   ___ a few thousand atoms or molecules
   ___ a few hundred atoms or molecules
   ___ one atom or molecule
   ___ 6.022×10^{23} atoms or molecules
2. Nanotechnology is the science of matter ______ in size.
   ___ 10^{-2} m
   ___ 10^{-3} m
   ___ 10^{-6} m
   ___ 10^{-9} m
3. Nanotechnology is a discipline that borrows concepts from _______ to explain the unexpected behaviors of nanoscale materials.
   ___ physics
   ___ chemistry
   ___ materials science
   ___ physics, chemistry, and materials science
4. The first integrated circuits (ICs) contained feature sizes no smaller than _____.

___ 50 μm
___ 5 μm
___ 100 nm
___ 10 nm

5. ICs in the early 1970s contained:
___ 2300 transistors
___ 15,320 transistors
___ 525,000 transistors
___ 1,256,000 transistors

6. The Lycurgus cup dates back to the _____ century.
___ 4th
___ 6th
___ 8th
___ 10th

7. Nanoparticles were discovered in the glass of the Lycurgus cup using:
___ an atomic force microscope
___ a scanning electron microscope
___ a scanning tunneling microscope
___ a transmission electron microscope

8. The diameter of the nanoparticles in the glass of the Lycurgus cup is ____ nm.
___ 70
___ 50
___ 30
___ 10

9. The Lycurgus cup appears _____ when light is reflected from its surface and _____ when light shines through the cup.
___ green, red
___ red, green
___ yellow, green
___ blue, yellow

10. Colloidal _____ and _____ nanoparticles are found in the Lycurgus cup making it possible for the cup to appear two different colors when light passes through and is reflected off the cup.
___ gold, lead
___ iron, copper

___ silver, gold
___ nickel, tin

11. The _____ component of the Lycurgus cup is responsible for the reddish color and the _____ component is responsible for the greenish color.
    ___ gold, copper
    ___ gold, silver
    ___ copper, silver
    ___ lead, copper
12. _____ is/are another example of a historical artifact containing metal nanoparticles.
    ___ swords
    ___ furniture
    ___ clothing
    ___ stained glass
13. Swords/saber blades have increased strength due to the incorporation of what type of nanomaterial?
    ___ metal nanoparticles
    ___ quantum dots
    ___ carbon nanoparticles
    ___ all of the above
14. The nanomaterials responsible for providing Damascus Swords with high mechanical strengths were created using:
    ___ heat only
    ___ organic (carbon containing) debris only
    ___ hammering only
    ___ heat, hammering, organic debris
15. Medieval artisans created the ruby red color in stained glass windows by mixing what compound with molten glass?
    ___ silver nitrate
    ___ gold chloride
    ___ copper sulfate
    ___ barium nitrate
16. In medieval stained glass, the nanoparticles have diameters of ____ nm.
    ___ 30
    ___ 25
    ___ 20
    ___ 15

17. A ____ is a mixture that contains two phases of matter, generally solids dispersed in a liquid or gas medium.
    ___ solution
    ___ suspension
    ___ colloid
    ___ emulsion
18. Michael Faraday discovered that ruby gold demonstrated what property?
    ___ ruby gold polarizes light
    ___ ruby gold produces different-colored solutions
    ___ ruby gold is more reactive than bulk gold
    ___ all of the above
19. Evidence of Michael Faraday's synthesis of gold nanoparticles was observed when the solution of sodium tetrachloroaurate ($NaAuCl_4$) turned from ______ to ______ after the addition of phosphorus.
    ___ green, blue
    ___ blue, green
    ___ red, yellow
    ___ yellow, red
20. Michael Faraday's research with Ruby Gold led to this conclusion:
    ___ ruby gold gold is composed of dissolved ions
    ___ carbon nanotubes exist in the soot produced by flames
    ___ matter can be shaped one at a time
    ___ fine materials (nanomaterials) have properties which are different from bulk materials
21. In 1959, Richard Feynman of the California Institute of Technology gave the first lecture on what topic?
    ___ integrated circuits
    ___ metal nanoparticles
    ___ carbon nanotubes
    ___ atom by atom assembly
22. In his lecture entitled "There's plenty of room at the bottom," Richard Feynman suggested that the wires found in computing devices would have diameters _____ in size.
    ___ 10 or 100 μm
    ___ 10 or 100 nm
    ___ 10 or 100 atoms
    ___ 10 or 100 cm

23. In his book *Engines of Creation*, K. Eric Drexler believed that atomic and molecular scale manipulation was possible because _____ were still responsible for holding atoms together.

___ magnetism

___ electrostatic forces

___ chemical bonds

___ magnetism and electrostatic forces

24. K. Eric Drexler believed that molecular assembly is driven by:

___ chemical bonding

___ electrostatic forces

___ chemical reactions

___ magnetism

25. K. Eric Drexler believed biochemistry would be able to use _____ to create motors, bearings, and moving parts, which would be able to handle individual molecules.

___ DNA

___ proteins

___ mitochondria

___ cells

26. In 1989, Don Eigler of IBM achieved atomic scale manipulation of xenon atoms using:

___ a scanning electron microscope

___ a transmission electron microscope

___ an atomic force microscope

___ a scanning tunneling microscope

27. The bright colors found in some insects is due to:

___ micro- and nanoscale molecular pigments

___ micro- and nanoscale structures arranged in a periodic fashion

___ nanoparticles

___ carbon nanotubes

28. Due to _______ exerted by the micro- and nanoscale structures on their toes, Geckos can climb and run on most surfaces.

___ dipole forces

___ hydrogen bonding

___ London dispersion forces

___ electrostatic forces

29. Phenomena associated with water beading up and rolling off a surface without wetting it is referred to as:
    ___ hydrophilicity
    ___ hydrophobicity
    ___ hydrogen bonding
    ___ chemical bonding

References

Aitken, R. J., M. Q. Chaudhry, A. B. A. Boxall, and M. Hull. 2006. "Manufacture and use of nanomaterials: Current status in the UK and global trends." *Occup. Med.* 56: 300–306.

Autumn, K., P. H. Niewiarowski, and J. B. Puthoff. 2014. "Gecko adhesion as a model system for integrative biology, interdisciplinary science, and bioinspired engineering." *Annu. Rev. Ecol. Evol. Syst.* 45: 445–470.

Autumn, K., M. Sitti, Y. A. Liang, A. M. Peattie, W. R. Hansen, S. Spoonberg, T. W. Kenny, R. Fearing, J. N. Isrealachvili, and R. J. Full. 2002. "Evidence for van der Waals adhesion in gecko setae." *Proc. Natl Acad. Sci. USA* 99: 12252–12256.

Baird, D., and A. Shew. 2004. "Probing the history of scanning tunneling microscopy." In *Discovering the Nanoscale*, edited by D. Baird, A. Nordmann and J. Schummer, 145–156. Amsterdam: IOS Press.

Balani, K., R. G. Batista, D. Lahiri, and A. Agarwal. 2009. "The hydrophobicity of a lotus leaf: A nanomechanical and computational approach." *Nanotechnology* 20: 305707.

Ball, P. 2009. "Feynman's Fancy." *Chemistry World*, January: 58–62.

Browne, M. W. 1990. "2 Researchers Spell 'I.B.M.,' Atom by Atom." *The New York Times*, April 5.

Chang, K. 2005. "Tiny Is Beautiful: Translating 'Nano' Into Practical." *The New York Times*, February 22.

Drexler, K. E. 1986. *Engines of Creation The Coming Era of Nanotechnology*. New York: Anchor Books.

El-Aawar, H. 2015. "Increasing the transistor count by constructing a two-layer crystal square on a single chip." *Int. J. Comput. Sci. Inf. Technol. Adv. Res.* 7: 97–105.

Freestone, I., N. Meeks, M. Sax, and C. Higgitt. 2007. "The Lycurgus cup - a roman nanotechnology." *Gold Bull.* 40: 270–277.

Hong, S. H., J. Hwang, and H. Lee. 2009. "Replication of cicada wing's nano-patterns by hot embossing and UV nanoimprinting." *Nanotechnology* 20: 385303.

Horikoshi, S., and N. Serpone. 2013. "Introduction to nanoparticles." In *Microwaves in Nanoparticle Synthesis Fundamentals and Applications*, edited by S. Horikoshi and N. Serpone, 1–23. Weinheim: Wiley-VCH.

Hsiung, B. K., T. A. Blackledge, and M. D. Shawkey. 2014. "Structural color and its interaction with other color-producing elements: Perspectives from spiders." Edited by R. Liang and J. A. Shaw. *The Nature of Light: Light in Nature V. San Diego: Proc. of SPIE 9187.* 91870B-1 - 91870B-20.

Kaiser, W., and P. Kuerz. 2008. "EUVL—extreme ultraviolet lithography." *Optik & Photonik* 3: 35–39.
Ko, D. H., J. R. Tumbleston, K. J. Henderson, L. E. Euliss, J. M. DeSimone, R. Lopez, and E. T. Samulski. 2011. "Biomimetic microlens array with antireflective "moth-eye" surface." *Soft Matter* 7: 6404–6407.
Kolle, M., P. M. Salgard-Cunha, M. R. J. Scherer, F. Huang, P. Vukusic, S. Mahajan, J. J. Baumberg, and U. Steiner. 2010. "Mimicking the colourful wing scale structure of the Papilio blumei butterfly." *Nat. Nanotechnol.* 5: 511–515.
Kolwas, K., A. Derkachova, and D. Jakubczyk. 2016. "Tailoring plasmon resonances in metal nanospheres for optical diagnostics of molecules and cells." In *Nanomedicine and Tissue Engineering State of the Art and Recent Trends*, edited by R. Augustine, N. Kalarikkal, O. S. Oluwafemi, K. S. Joshy and S. Thomas, 141–182. Boca Raton: CRC Press/Taylor and Francis Group.
Madou, M. J. 2011. *Fundamentals of Microfabrication and Nanotechnology - Manufacturing Techniques for Microfabrication and Nanotechnology*. Vol. 11. Boca Raton: CRC Press/Taylor & Francis Group.
Mulvaney, P. 2015. "Nanoscience vs nanotechnology—defining the field." *ACS Nano* 9: 2215–2217.
Pradeep, T. 2007. *Nano: The Essentials Understanding Nanoscience and Nanotechnology*. New Delhi: Tata McGraw-Hill.
Rogers, B., J. Adams, and S. Pennathur. 2013. *Nanotechnology The Whole Story*. Boca Raton: CRC Press/Taylor and Francis Group.
Roukes, M. 2001. "Plenty of Room, Indeed." *Scientific American*, September: 48–57.
Sitti, M., and R. S. Fearing. 2003. "Synthetic gecko foot-hair micro/nano-structures as dry adhesives." *J. Adhes. Sci. Technol.* 18: 1055–1074.
Srinivasan, C. 2007. "Do Damascus swords reveal India's mastery of nanotechnology?" *Curr. Sci.* 92: 279–280.
Thompson, D. 2007. "Michael Faraday's recognition of ruby gold: The birth of modern nanotechnology." *Gold Bull.* 40: 267–269.
Verbanic, S., O. Brady, A. Sanda, C. Gustafson, and Z. J. Donhauser. 2014. "A novel general chemistry laboratory: Creation of biomimetic superhydrophobic surfaces through replica molding." *J. Chem. Educ.* 91: 1477–1480.
Zhang, Y., Y. R. Leu, R. J. Aitken, and M. Riediker. 2015. "Inventory of Engineered nanoparticle-containing consumer products available in the Singapore retail market and likelihood of release into the aquatic environment." *Int. J. Environ. Res. Public Health* 12: 8717–8743.

2

Chemistry Foundations in Nanotechnology

Key Objectives

- Know the difference between atoms and ions
- Know the difference between oxidation and reduction
- Become familiar with the different types of chemical reactions
- Become familiar with the different types of chemical bonds
- Know various types of intermolecular forces and their roles in nanotechnology

2.1 Introduction

Chemistry has been the driving force behind many technological advances (Pradeep 2007). Products with sizes from 0.1 to 10 nm can be created by breaking and forming bonds between atoms or groups of atoms. New chemical reaction methodologies have led to the production of uniform nanostructures with various shapes (spheres, rods, wires, cubes) and compositions (organics, metals, oxides, and semiconductors) (Love et al. 2005).

2.2 Atoms

Imagine an apple the size of the Earth; in comparison, a normal-sized apple becomes the size of an atom next to the Earth-sized apple (Rogers, Adams and Pennathur 2013). The basic model of an atom consists of a dense, positively charged nucleus surrounded by negatively charged electrons orbiting the nucleus, as shown in Figure 2.1. The nucleus consists of protons and neutrons and makes up the majority of the atom's mass (Ratner and Ratner 2003). Atoms

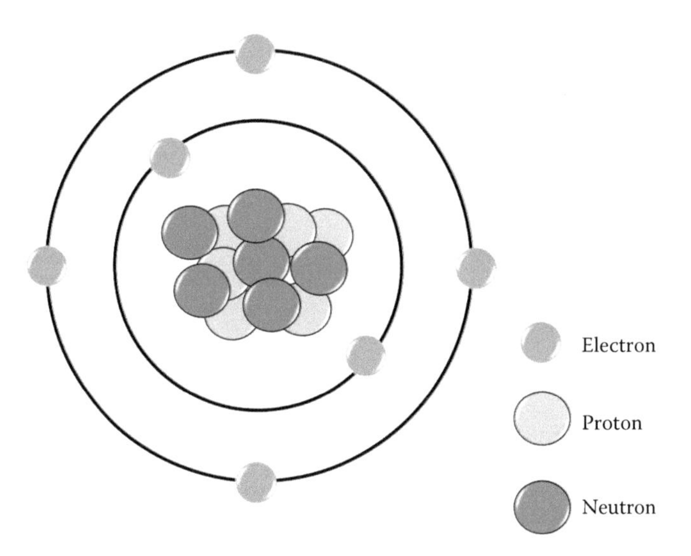

FIGURE 2.1
Basic model of the atom.

have a specific quantity of protons known as the atomic number. Atoms have no net electrical charge because the number of electrons exactly balances the charge of the nucleus, so there is one electron for every proton (Ratner and Ratner 2003). All atoms are approximately 0.1 nanometer in diameter (Ratner and Ratner 2003).

2.3 Ions

During the course of a chemical reaction, the nucleus of an atom remains unchanged, however, electrons can be gained or lost. When atoms gain or lose electrons, a charged particle called an ion is formed. When a metal atom loses electrons, an ion with a positive charge, a cation, is formed. A negatively charged ion, an anion, is formed when nonmetal atoms gain electrons. The sodium atom, for instance, has 11 protons and 11 electrons. Sodium atoms easily lose one electron, resulting in the formation of a cation having 11 protons and 10 electrons. The resulting ion has a net charge of 1^+ (Figure 2.2a). Chlorine atoms have 17 protons and 17 electrons. Chlorine atoms gain one electron, forming an anion with 17 protons and 18 electrons with a net charge of 1^- (Figure 2.2b).

Generally, metal atoms lose electrons and nonmetal atoms gain electrons (Brown et al. 2012). Positive ions are smaller than neutral atoms because there are fewer electrons, and they are more closely held by the positive charge of the nucleus. Negative ions are larger than the neutral atoms due to the extra electrons added to the electron cloud (Ratner and Ratner 2003).

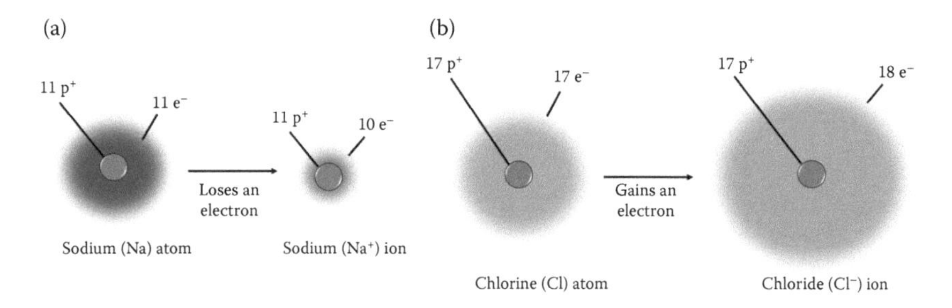

FIGURE 2.2
Formation of sodium ions (a) and chlorine ions (b).

2.3.1 Oxidation and Reduction

Often, ions are used as the building blocks for nanomaterials. Nanomaterial synthesis often involves a chemical reaction shuttling electrons between particles. Oxidation occurs when an atom or ion loses electrons. Reduction occurs when an atom or ion gains electrons. A particle experiencing oxidation is called the reducing agent, and one experiencing reduction is the oxidizing agent (Brown et al. 2012). Both processes are outlined in Figure 2.2a–b. In Chapter 6, a procedure will be described using reduction to synthesize metal nanoparticles and nanowires.

2.4 Subatomic Particles

In the mid-1800s, J. J. Thomson studied electrical discharges in a glass tube containing a partial vacuum. These experiments led to the discovery of electrons. Thomson applied high voltages to the electrodes at the ends of the glass tube; as a result, radiation was observed between the electrodes. This radiation was named cathode rays because they originated at the negative electrode (cathode) and traveled to the positive electrode (anode) (Figure 2.3). Cathode rays consist of negatively charged electrons, 2000 times lighter than hydrogen, the smallest atom (Ratner and Ratner 2003).

The gold foil experiment (Figure 2.4), conducted by Ernest Rutherford, led to the discovery of the atomic nucleus. This experiment involved firing alpha particles towards thin gold foil. Most of the alpha particles passed through the foil without being scattered; however, when alpha particles came close to a gold nucleus, the positive charge of the gold nucleus deflected the positively charged alpha particles, which then struck a fluorescent screen. From this experiment, Rutherford concluded that most of an atom's mass resides in a dense region called the nucleus. Additionally, it was determined that most of the volume of an atom is empty space. Additional experiments conducted during subsequent years led to the discovery of positive particles (protons) and neutral particles (neutrons) in the nucleus (Brown et al. 2012).

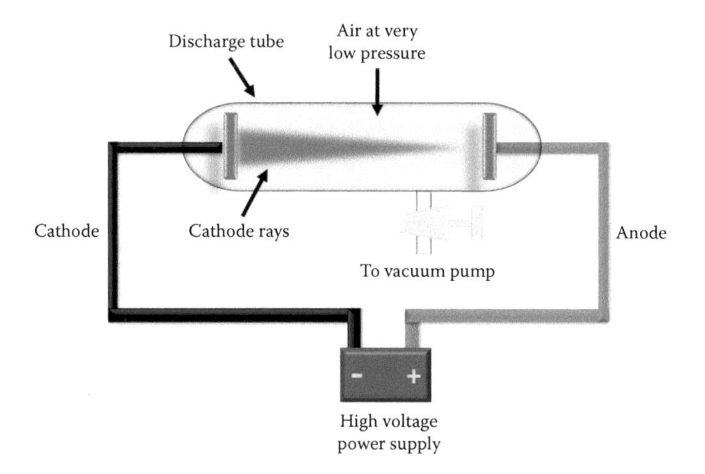

FIGURE 2.3
Cathode ray tube.

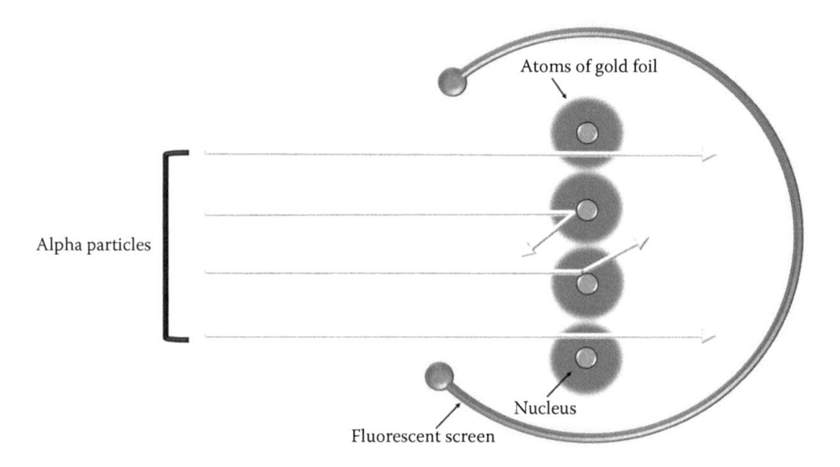

FIGURE 2.4
The gold foil experiment.

2.5 Chemical Bonding

Atoms join together to form molecules, the smallest part of a compound. Reports state that approximately 30 atoms bonded together form a molecule one nanometer in size. Chemical bonds are important to nanotechnology because they can behave as hinges, bearings, or other types of nanoscale mechanical devices. One category of bonding, ionic bonding, occurs between metal atoms (found on the left side of the periodic table) and nonmetal atoms

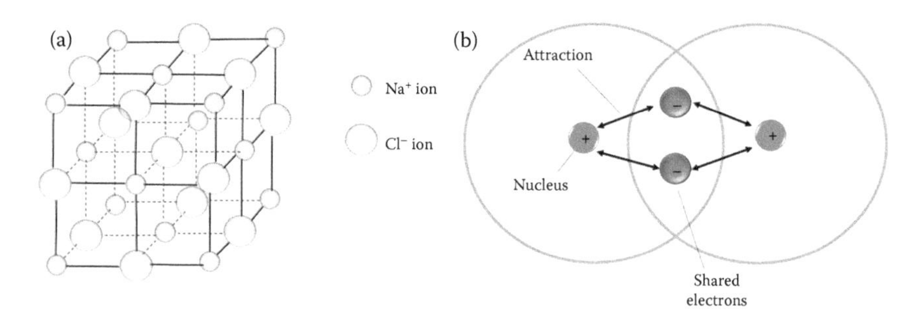

FIGURE 2.5
Crystal lattice structure for sodium chloride (a). Formation of a covalent bond (b).

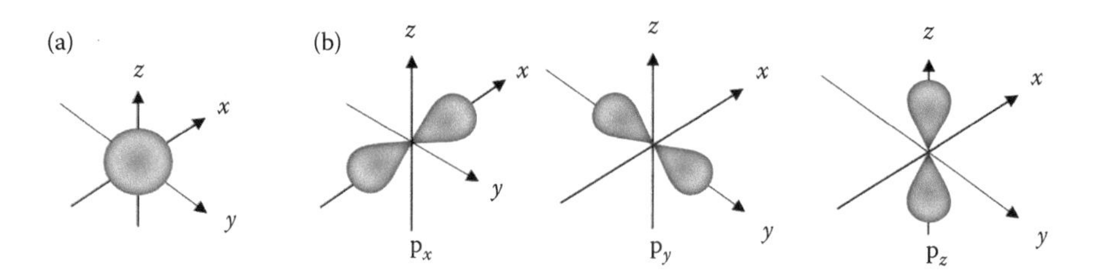

FIGURE 2.6
s (a) and p (b) atomic orbitals.

(found on the right side of the periodic table). Compounds with ionic bonds are brittle and have high melting points. The ions forming these compounds arrange themselves in a crystal lattice structure, as shown in Figure 2.5a. Ionic bonding involves electrostatic forces that maintain ions in a rigid, periodic, three-dimensional structure (Figure 2.6a). Covalent bonds are formed when nonmetal atoms share electron pairs. When two nonmetal atoms are close to each other, the positively charged nuclei repel each other and the negatively charged electrons repel each other. However, the nuclei and electrons attract each other, resulting in a concentration of negative charge between the atoms (Figure 2.5b). The shared pair of electrons in any covalent bond behaves as a "glue," which holds atoms together (Ratner and Ratner 2003). Compounds containing covalent bonds tend to be gases, liquids, or solids with low melting points (Ratner and Ratner 2003).

2.6 Chemical Reactions

Chemical reactions are often used to create various forms of nanomaterials. Chemical equations are used to represent chemical reactions. Examples of chemical equations are shown in Table 2.1. The substances to the left

TABLE 2.1
Examples of Chemical Equations

Reaction Type	Format	Example
Synthesis reaction	$A + B \rightarrow AB$	$N_2 + 3H_2 \rightarrow 2NH_3$
Decomposition reaction	$AB \rightarrow A + B$	$CuCO_3 \rightarrow CuO + CO_2$
Single replacement reaction	$A + BC \rightarrow B + AC$	$Cu + 2AgNO_3 \rightarrow 2Ag + Cu(NO_3)_2$
Double replacement reaction	$AB + CD \rightarrow AD + BC$	$2KI + Pb(NO_3)_2 \rightarrow PbI_2 + 2KNO_3$

of the arrow are the reactants, and those to the right of the arrow are the products. The arrow is interpreted as "yields." Numbers next to the formulas are known as coefficients. They indicate the relative numbers of molecules of each substance involved in the reaction. Atoms are neither created nor destroyed in chemical reactions, and for this reason, a chemical equation must have equal numbers of atoms of each element on each side of the arrow resulting in a balanced equation. The smallest whole-number coefficients are used to balance the equation (Ratner and Ratner 2003).

2.7 Quantum Mechanics

Rules associated with classical physics no longer apply at the nanoscale. For instance, electrons show properties of both waves and particles. Quantum mechanics was developed to describe the unique behaviors of electrons (Wilson et al. 2002). Quantum mechanics was developed by Erwin Schrodinger. This field involves determining the probability of finding an electron in a region of space (Brown et al. 2012). In quantum mechanics, the properties of particles such as electrons and photons are described using wave functions (Ψ). Furthermore, the square of the wave function (Ψ^2) is proportional to the probability of finding a particle in a specific region of space. Quantum mechanical properties are exploited in instruments used to study the atomic scale. For example, the scanning tunneling microscope (STM) allows users to visualize atoms due to a quantum effect known as tunneling (Baird and Shew 2004).

2.8 Atomic Orbitals

Atomic orbitals are derived from wave functions that describe the probable spatial distribution of electrons (Brown et al. 2012). The "s" orbital and the

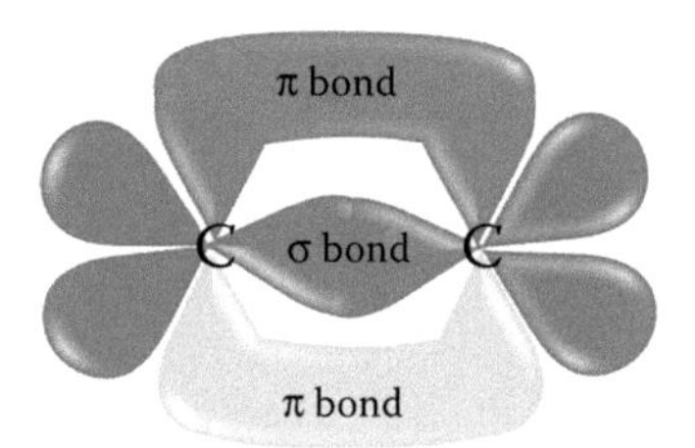

FIGURE 2.7
Sigma (σ) and pi (π) bonds.

"p" orbitals are shown in Figures 2.6a and 2.6b, respectively. Orbitals play an important role in chemical bond formation. Ionic bonds are formed when electrons are lost from an orbital in the valence shell of metal atoms and fill empty orbitals in the valence shells of nonmetal atoms.

Covalent bonds form when nonmetal atoms share electron pairs via atomic orbital overlap. In covalently bonded materials, atomic orbitals can overlap one of two ways. Atomic orbitals can overlap directly forming sigma (σ) bonds, while sideways orbital overlap can occur forming pi (π) bonds, both of which are shown in Figure 2.7. The presence of π bonds helps to explain why carbon-based nanomaterials, such as carbon nanotubes and graphene, exhibit high electrical conductivity, as explained in Chapter 4.

2.9 Molecular Orbitals

Molecular orbitals can be used to understand the basic operation of molecular electronics, which will be discussed in Chapter 5. Molecular orbitals have characteristics similar to atomic orbitals. For instance, molecular orbitals can hold a maximum of two electrons. Additional similarities include electrons occupying the same molecular orbital require opposite spins, and molecular orbitals have definite energies. Whenever two atomic orbitals overlap, two molecular orbitals are formed: one bonding, one antibonding, as shown in Figure 2.8 (Miessler, Fischer and Tar 2014).

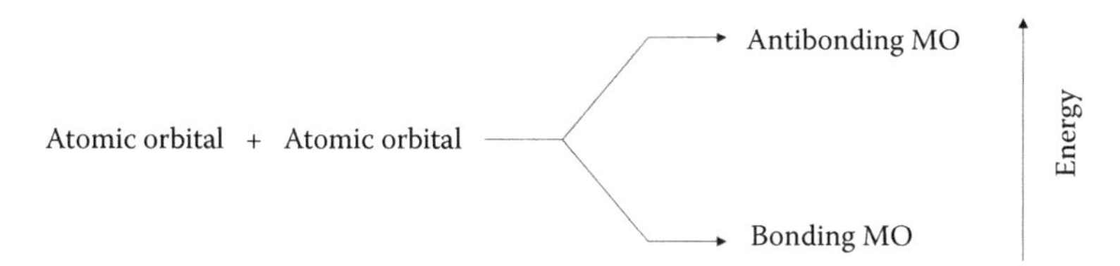

FIGURE 2.8
Molecular orbital diagram.

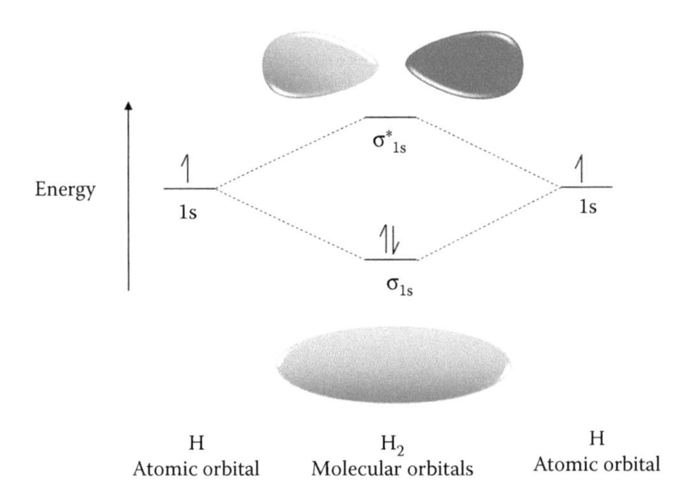

FIGURE 2.9
Molecular orbital diagram for the formation of hydrogen gas (H_2).

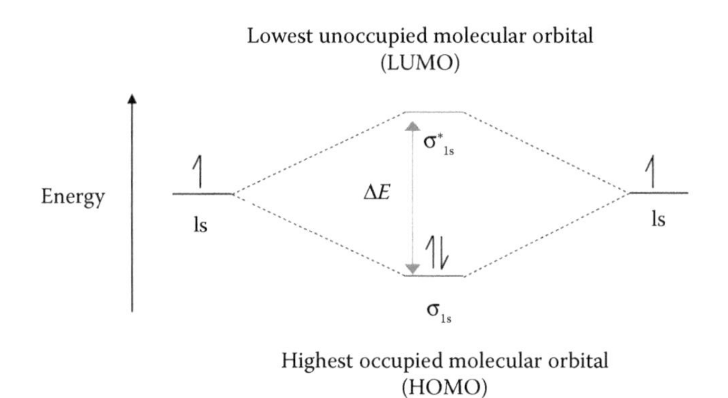

FIGURE 2.10
The energy difference (ΔE) between the HOMO and the LUMO. When sufficient energy is added to the molecule, an electron can be promoted from the HOMO to the LUMO.

Bonding orbitals are constructive combinations of atomic orbitals. Antibonding orbitals are destructive combinations of atomic orbitals. They have a nodal plane where electron density equals zero, as illustrated in the molecular orbital diagram for hydrogen gas (H_2) shown in Figure 2.9 (Miessler, Fischer and Tar 2014).

Molecular orbitals are involved in chemical reactions. These orbitals are referred to as the highest occupied molecular orbital (HOMO) and the lowest unoccupied molecular orbital (LUMO) (Figure 2.10) (Miessler, Fischer and Tar 2014). Electrons in the HOMO are the most loosely held electrons in the molecule and participate in chemical bond formation. The LUMO receives electrons during a chemical reaction (Miessler, Fischer and Tar 2014).

2.10 Intermolecular Forces

Forces that exist between molecules are referred to as intermolecular forces. It is important to consider these forces when investigation nanotechnology. Often, these forces drive the assembly of nanomaterials. These forces require less energy to disrupt them, compared to chemical bonds. Disrupting or encouraging intermolecular forces is responsible for phase changes, however, when a molecular substance changes phase (i.e., liquid to gas), the molecules remain intact. Physical properties of liquids (boiling points, melting points, etc.) reflect the strength of the intermolecular forces. For instance, to convert a liquid to a gas, molecules of the liquid must overcome their attractive forces to separate and form a vapor. The stronger the attractive forces, the higher the temperature required to change the phase, thus the liquid boils at a higher temperature. Similarly, melting points increase as the strengths of the intermolecular forces increase. London dispersion forces, dipole–dipole attractions, and hydrogen bonding are the three types of intermolecular attractions existing between molecules (Brown et al. 2012).

2.10.1 London Dispersion Forces

London dispersion forces are attractions between atoms and molecules with temporary, instantaneous dipoles. Temporary dipoles are formed when, at any given instant, electrons congregate on one side of the atom or molecule, leading to the formation of instantaneous or temporary regions of charge. At that point, the atom or molecule becomes temporarily polarized. Temporary dipoles in one atom or molecule can induce temporary dipoles in an adjacent particle resulting in an attraction between the two (Figure 2.11). London dispersion forces are short-range forces. They are significant when atoms aand molecules are very close together. At the nanoscale, these forces are dominant (Rogers, Adams and Pennathur 2013).

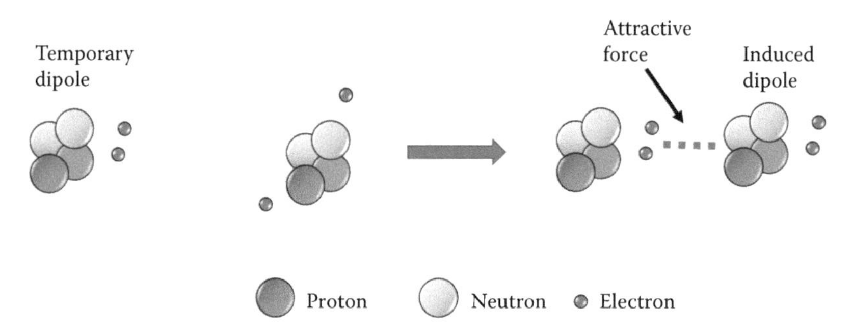

FIGURE 2.11
Formation of London dispersion forces.

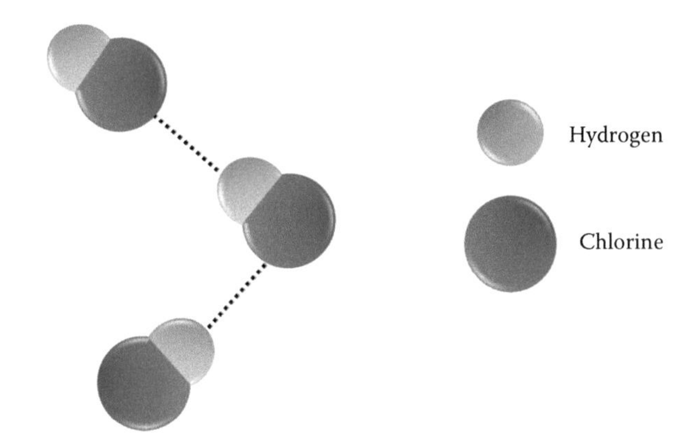

FIGURE 2.12
Dipole forces acting between hydrogen chloride (HCl) molecules.

2.10.2 Dipole–Dipole Interactions

The presence of permanent dipoles in polar molecules results in dipole–dipole interactions between these molecules. These forces occur due to the attractions between the partially positive end of one molecule and partially negative end of a neighboring molecule (Figure 2.12). To add, repulsions can also occur when the positive (or negative) ends of two molecules are facing each other.

2.10.3 Hydrogen Bonding

Hydrogen bonding is an intermolecular attraction between the hydrogen atom in one molecule, and a nearby atom of fluorine, oxygen, or nitrogen in an adjacent molecule (Figure 2.13). Nitrogen, oxygen, and fluorine are very

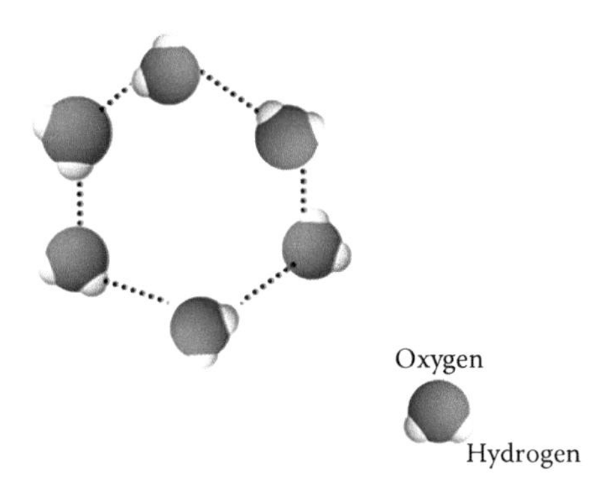

FIGURE 2.13
Hydrogen bonding between water molecules.

electronegative, and due to the small size of the electron poor hydrogen atom it can make a close approach with an electronegative atom and interact strongly with it (Brown et al. 2012). It is interesting to point out that hydrogen bonding maintains water in liquid form over a larger temperature range than other molecules similar in size (Rogers, Adams and Pennathur 2013).

2.11 Polymers

Polymers are long chain molecules often used in nanomaterial synthesis as precursors (Lu, Shah and Xu 2017), templates (Sanders et al. 2017), or directing agents (Zhu et al. 2011). Polymers consist of long chains of atoms connected together using covalent bonds. Adjacent polymer chains are held together by weak intermolecular forces (Brown et al. 2012). There are several naturally occurring substances that exist as polymers, such as starch, cellulose, and proteins; these materials are found in both plants and animals (Brown et al. 2012). Jons Jakob Berzelius introduced the word polymer in 1827, which is derived from the Greek words *polys* (many) and *meros* (parts). Polymers are molecules formed via *polymerization* (joining together) of smaller molecules referred to as monomers. An example of a commonly used polymer in pipes, polyvinyl chloride (PVC), and the corresponding monomer, vinyl chloride, are shown in Figure 2.14 (Brown

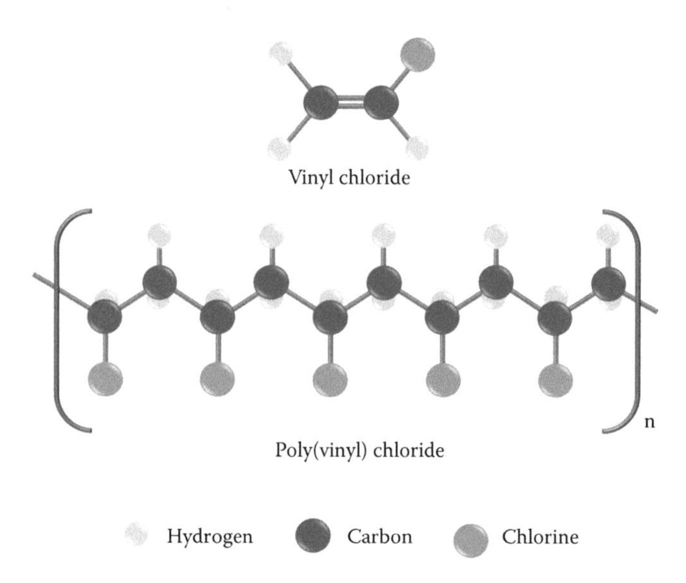

FIGURE 2.14
Poly(vinyl chloride) and the corresponding monomer, vinyl chloride.

et al. 2012). Polymers are often referred to as plastics, and these materials can be shaped using heat and pressure (Brown et al. 2012). Thermoplastics are particles that can be reshaped with heat. Additionally, thermoplastics can be melted down and recycled for other uses. Conversely, a thermoset is a polymeric material shaped using a chemical process and cannot be reshaped readily. Elastomers are polymers exhibiting a rubbery or elastic behavior. When elastomers are stretched or bent, these materials revert to their original shape upon removal of the applied force (Brown et al. 2012). It is important to point out that polymer chains are usually not straight because atoms are relatively free to rotate around the bonds between carbon atoms, resulting in flexibility (Brown et al. 2012). In addition to being flexible, polymers can possess regions exhibiting crystallinity. Crystallinity involves the alignment of polymer chains in regular arrays due to stretching or pulling the polymer chains. Intermolecular forces acting between polymer chains can also result in polymers that are denser, harder, less soluble, and more resistant to heat (Brown et al. 2012). In addition, polymers can be made stiffer by utilizing chemical bonds between chains. This process is referred to as cross-linking. The greater the number of cross-links, the more rigid the polymer (Brown et al. 2012).

2.12 Semiconductor Materials

Semiconductor materials have both metallic and nonmetallic characteristics. However, intentionally introducing impurities to levels as low as 0.01% can change the electrical resistance of a semiconductor 10,000-fold. Switching between the metallic and insulating characters of semiconductors is the basis of the transistor, an electronic switch in modern electronics. Semiconductors contain electrons and holes responsible for carrying charges through semiconductor materials. The type of charge carriers in the semiconductor can be varied by controlling the type and concentration of dopants added to the material. In pure silicon, all four electrons in the valence shell are shared with neighboring silicon atoms. Since there are no free electrons, silicon does not intrinsically conduct electricity. To change the conductivity of silicon minute amounts of a material, such as antimony (n-type dopant), is added to the silicon crystal. Antimony atoms provide extra electrons. Additionally, silicon can be doped with elements having at least one less electron than the host material, such as boron. This results in the formation of p-type semiconductors. The missing electron creates a positively charged "hole," which can also carry a current (Wilson et al. 2002). Examples of p and n-type doping of silicon is illustrated in Figure 2.15.

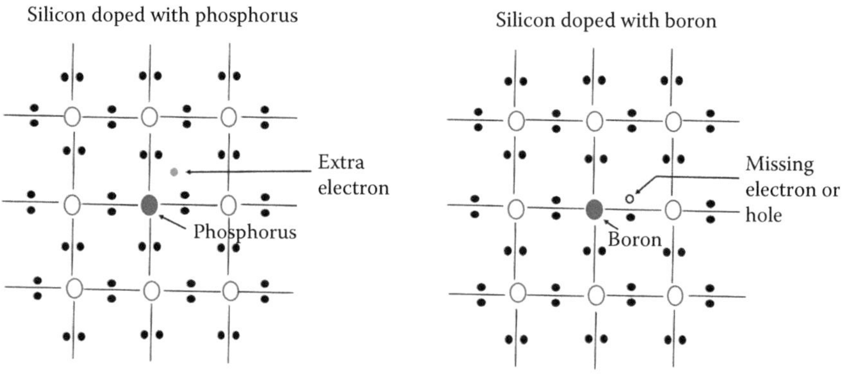

Silicon doped with phosphorus
Silicon doped with boron

Extra electron
Phosphorus
Missing electron or hole
Boron

FIGURE 2.15
Silicon doping with P and N dopants.

2.13 End-of-Chapter Questions

1. A/an _____ is the smallest part of an element.
   ___ molecule
   ___ atom
2. A/an _____ is the smallest part of a compound.
   ___ molecule
   ___ atom
3. Breaking and forming bonds between atoms or groups of atoms forms molecular materials with sizes ranging from:
   ___ 0.1 mm to 100 mm
   ___ 0.1 cm to 100 cm
   ___ 0.1 μm to 100 μm
   ___ 0.1 nm to 100 nm
4. Atoms are approximately _____ nm in diameter:
   ___ 0.1
   ___ 0.01
   ___ 0.001
   ___ 0.0001
5. When metal atoms lose electrons _____ are formed.
   ___ cations
   ___ anions

6. When nonmetal atoms gain electrons ___ are formed.

 ___ cations

 ___ anions

7. If an atom or ion loses electrons, ______ occurs.

 ___ oxidation

 ___ reduction

8. If an atom or ion gains electrons, _____ reduction occurs.

 ___ oxidation

 ___ reduction

9. The addition of impurities to a semiconductor can change its electrical resistance _____ fold.

 ___ 10

 ___ 100

 ___ 1000

 ___ 10,000

10. Extra electrons are provided by _____ dopants.

 ___ n-type

 ___ p-type

11. Positively charged holes are provided by _____ dopants.

 ___ n-type

 ___ p-type

12. The bonds that form between metal and nonmetal atoms are _____.

 ___ covalent bonds

 ___ ionic bonds

13. The type of bonds that form when atoms share electron pairs are _____.

 ___ covalent bonds

 ___ ionic bonds

14. Numbers in front of formulas in chemical equations (coefficients) are used to indicate:

 ___ relative number of molecules involved in the reaction

 ___ temperature of the reactants and products

 ___ melting points of the reactants and products

 ___ boiling points of the reactants and products

15. A balanced chemical equation should contain the smallest possible whole-number coefficients.

 ___ true

 ___ false

16. Quantum mechanics is used to determine the probability of finding _____ in an atom.

 ___ neutron

 ___ proton

 ___ electron

 ___ electrons and protons

17. The _____ nature of electrons allows them to tunnel through insulating layers that normally block electron flow.

 ___ wave

 ___ particle

18. Which of the following is proportional to the probability of finding an electron or photon in a specific region of space?

 ___ Ψ

 ___ Ψ^2

19. Likely positions of electrons occupying a main energy level of atoms are referred to as:

 ___ energy levels

 ___ chemical bonds

 ___ nuclei

 ___ orbitals

20. Sigma bonds are formed from _____ overlap of atomic orbitals.

 ___ direct

 ___ sideways

21. Pi bonds are formed from _____ overlap of atomic orbitals.

 ___ direct

 ___ sideways

22. Molecular orbitals are formed when _____ overlap.

 ___ nuclei

 ___ atomic orbitals

 ___ nuclei and atomic orbitals

 ___ none of the above

23. Electrons in the _____ are the most loosely held electrons in a molecule.

 ___ HOMO

 ___ LUMO

24. Which of the following forces act between molecules with temporary, instantaneous dipoles?

 ___ dipole–dipole forces

 ___ hydrogen bonding

___ London dispersion forces
___ magnetism

25. Which of the following forces act between molecules with permanent dipoles?
    ___ dipole–dipole forces
    ___ hydrogen bonding
    ___ London dispersion forces
    ___ magnetism
26. Hydrogen bonding is an intermolecular force that acts between hydrogen and:
    ___ nitrogen
    ___ oxygen
    ___ fluorine
    ___ all of the above

References

Baird, D., and A. Shew. 2004. "Probing the history of scanning tunneling microscopy." In *Discovering the Nanoscale*, edited by D. Baird, A. Nordmann and J. Schummer, 145–156. Amsterdam: IOS Press.

Brown, T. L., H. E. LeMay, Jr., B. E. Bursten, C. J. Murphy, and P. M. Woodward. 2012. *Chemistry - The Central Science*. Boston: Prentice Hall.

Love, J. C., L. A. Estroff, J. K. Kriebel, R. G. Nuzzo, and G. M. Whitesides. 2005. "Self-assembled monolayers of thiolates on metals as a form of nanotechnology." *Chem. Rev.* 105: 1103–1169.

Lu, Y., K. W. Shah, and J. Xu. 2017. "Synthesis, morphologies and building applications of nanostructured polymers." *Polymers* 9: 506.

Miessler, G. L., P. J. Fischer, and D. A. Tar. 2014. *Inorganic Chemisry*. Boston: Pearson.

Pradeep, T. 2007. *Nano: The Essentials Understanding Nanoscience and Nanotechnology*. New Delhi: Tata McGraw-Hill.

Ratner, M., and D. Ratner. 2003. *Nanotechnology—A Gentle Introduction to the Next Big Idea*. Upper Saddle River: Prentice Hall.

Rogers, B., J. Adams, and S. Pennathur. 2013. *Nanotechnology The Whole Story*. Boca Raton: CRC Press/Taylor & Francis Group.

Sanders, W. C., R. Valcarce, P. Iles, J. S. Smith, G. Glass, J. Gomez, G. Johnson et al. 2017. "Printing silver nanogrids on glass." *J. Chem. Educ.* 94: 758–763.

Wilson, M., K. Kannangara, G. Smith, M. Simmons, and B. Raguse. 2002. *Nanotechnology—Basic Science and Emerging Applications*. Boca Raton: Chapman and Hall/CRC Press.

Zhu, J. J., C. X. Kan, J. G. Wan, M. Han, and G. H. Wang. 2011. "High-yield synthesis of uniform Ag nanowires with high aspect ratios by introducing the long-chain PVP in an improved polyol process." *J. Nanomater.* 2011: 1–7.

3

Physics Foundations in Nanotechnology

Key Objectives

- Gain familiarity with the properties of electromagnetic radiation
- Know the distinction between the wave nature and particle nature of light
- Become familiar with the band structures of insulators, semiconductors, and conductors
- Understand the role of the Fermi Level
- Understand the importance of the Bohr-exciton radius

3.1 Introduction

There are unique light–matter interactions occuring at the nanoscale. Nanoparticles with diameters less than 100 nm in size can be tuned to absorb specific wavelengths of light while reflecting other wavelengths. Even smaller nanoparticles, known as quantum dots, with diameters between 1 and 10 nm, can be tuned to emit specific wavelengths of light when exposed to ultraviolet radiation. To better understand the unique optical properties of nanoparticles described later in the text, an overview of relevant physics concepts is included in this chapter.

3.2 Electromagnetic Radiation

Visible light is just one type of electromagnetic radiation. Radio waves, infrared radiation, and X-rays are other examples of electromagnetic radiation (Brown et al. 2012). Electromagnetic radiation is energy travelling through space in the form of waves. All forms of electromagnetic radiation are listed in the

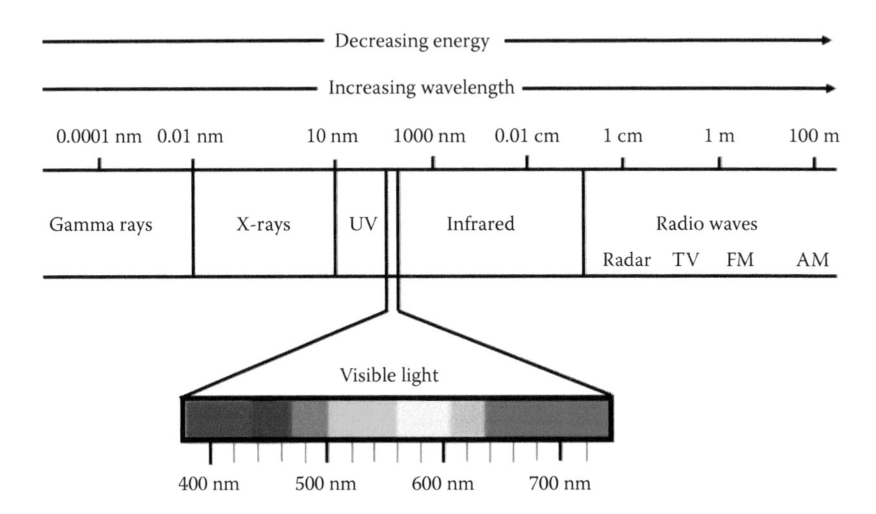

FIGURE 3.1
Electromagnetic spectrum.

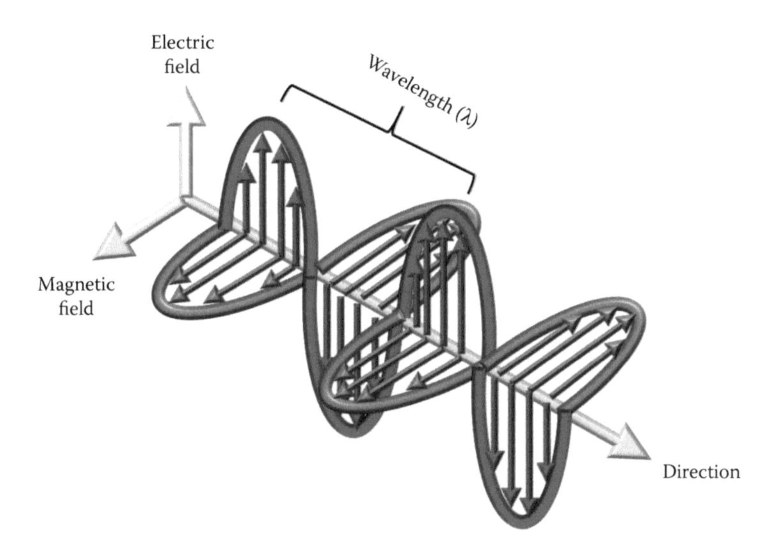

FIGURE 3.2
Propagation of electric and magnetic fields of electromagnetic radiation.

electromagnetic spectrum (Figure 3.1). This radiation consists of oscillating, perpendicular electric, and magnetic fields (Figure 3.2) (Brown et al. 2012).

Electromagnetic radiation can be released or absorbed by atoms in packets of energy. The energy, E, of electromagnetic radiation equals a constant times the frequency of the radiation, as shown in Equation 3.1:

$$E = h\nu \tag{3.1}$$

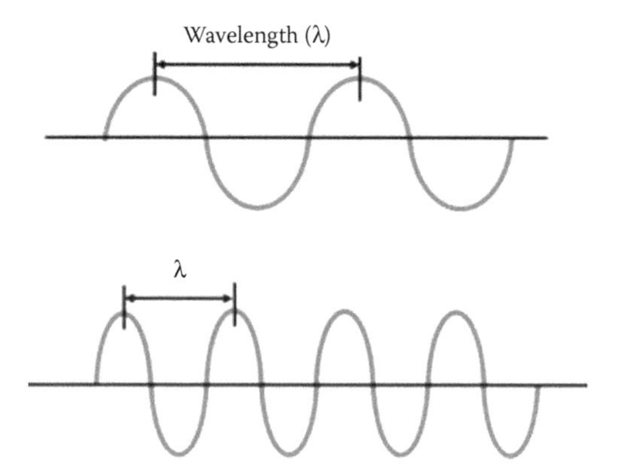

FIGURE 3.3
Relationship between frequency and wavelength.

The constant, *h*, is called Planck's constant and has a value 6.626×10^{-34} J-s (joule-second) (Serway, Vuille and Faughn 2006). All forms of electromagnetic radiation travel in a vacuum at the speed of light, *c*, with a value of 3.0×10^{8} m/s. The relationship between speed, frequency, and wavelength is shown in Equation 3.2:

$$c = \lambda \nu \tag{3.2}$$

The distance between two adjacent peaks (or two adjacent troughs) is called the wavelength (λ). The number of cycles that pass a given point each second is referred to as the frequency of the wave. Wavelength and frequency of electromagnetic radiation are inversely proportional. Long wavelengths have low frequencies, and short wavelengths have high frequencies (Figure 3.3). This inverse relationship between energy and wavelength can be expressed using Equation 3.3 (Brown et al. 2012). Common wavelengths associated with various forms of electromagnetic radiation are listed in Table 3.1.

$$E = \frac{hc}{\lambda} \tag{3.3}$$

TABLE 3.1
Wavelengths Associated with Electromagnetic Radiation

Unit	Symbol	Length (m)	Type of Radiation
Angstrom	Å	10^{-11}	X-ray
Nanometer	nm	10^{-9}	Ultraviolet, visible
Micrometer	μm	10^{-6}	Infrared
Millimeter	mm	10^{-3}	Microwave
Centimeter	cm	10^{-2}	Microwave
Meter	m	1	Television, radio
Kilometer	km	1000	Radio

3.3 The Wave Nature of Light

The wave nature of light was demonstrated by Thomas Young in 1801 using the double slit experiment. In this experiment, light was passed through two separate slits and projected on a screen (Serway, Vuille and Faughn 2006). The light emerging from the two slits generates a pattern with alternating bright and dark bands (Figure 3.4).

When light waves in phase arrive at a point on the screen, the waves combine, and constructive interference occurs (Figure 3.5a) producing bright

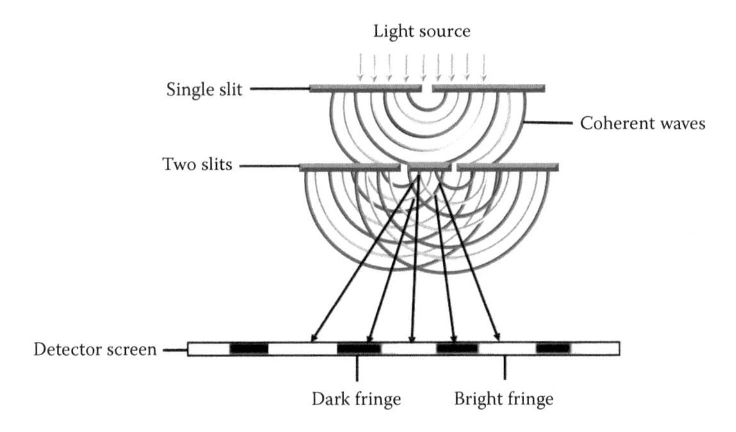

FIGURE 3.4
Double-slit experiment

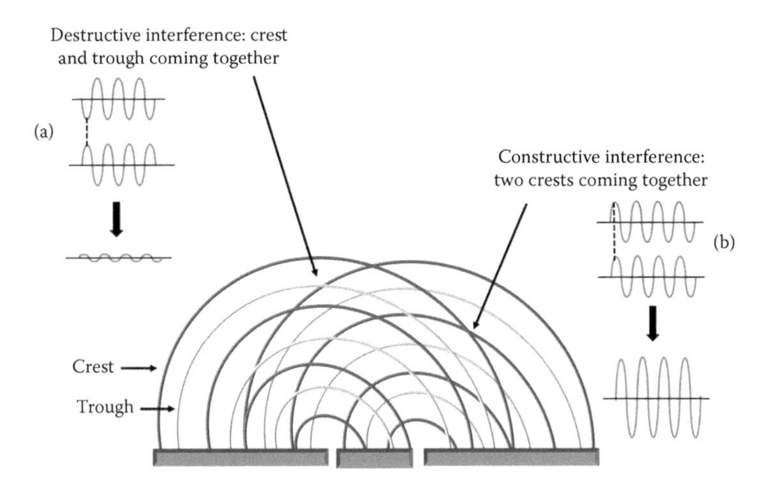

FIGURE 3.5
Destructive (a) and constructive (b) interference.

fringes. When out of phase light waves arrive at a point on the screen the waves cancel, and destructive interference occurs (Figure 3.5b) producing dark fringes.

3.4 Photoelectric Effect

Experimental evidence suggesting that light acts as a particle is observed with the photoelectric effect. This is observed when light possessing sufficient energy strikes a metal surface. When this happens, the metal emits electrons. Albert Einstein suggested that light striking the metal surface behaves like a stream of energy packets known as photons. Photons striking a metal surface transfers energy to electrons in the metal. In the dark, a current meter connected to a photocell reads zero. However, when the photocell is illuminated by light, a current is detected due to the emission of photoelectrons (Serway, Vuille and Faughn 2006) (Figure 3.6).

The photoelectric effect involves the work function (Φ), which is the energy required for electrons to overcome the attractive forces holding them in the metal. Photons striking the metal with less energy than the work function do not possess sufficient energy to release electrons from the metal; however, if photons possess energy greater than the work function, electrons are emitted. The photoelectric effect is currently used in digital cameras and burglar alarms (Rogers, Adams and Pennathur 2013).

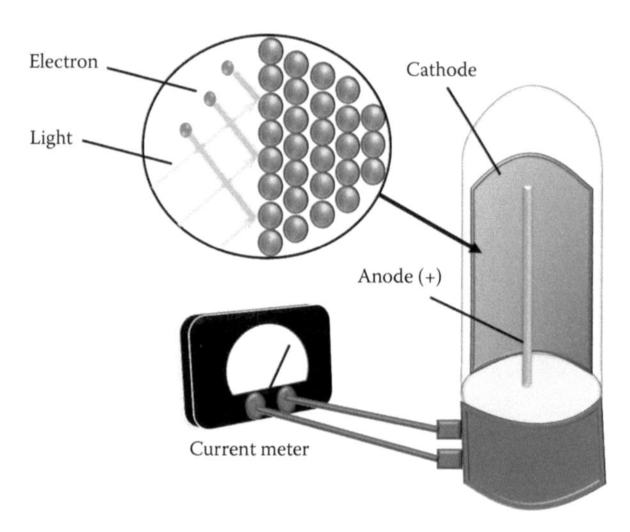

FIGURE 3.6
Photoelectric effect.

3.5 Band Structure

There is a specific order in which electrons are arranged in an atom, as shown in Figure 3.7. Orbitals, such as s and p, are the likely regions of electrons, as they reside on a quantized energy level. Orbitals can be filled with electrons or remain vacant.

Atomic orbitals overlap when atoms are near (Rogers, Adams and Pennathur 2013). Orbitals then reach a point where it is not possible to discern one orbital from the next. This results in the formation of energy bands (Figure 3.8) (Rogers, Adams and Pennathur 2013).

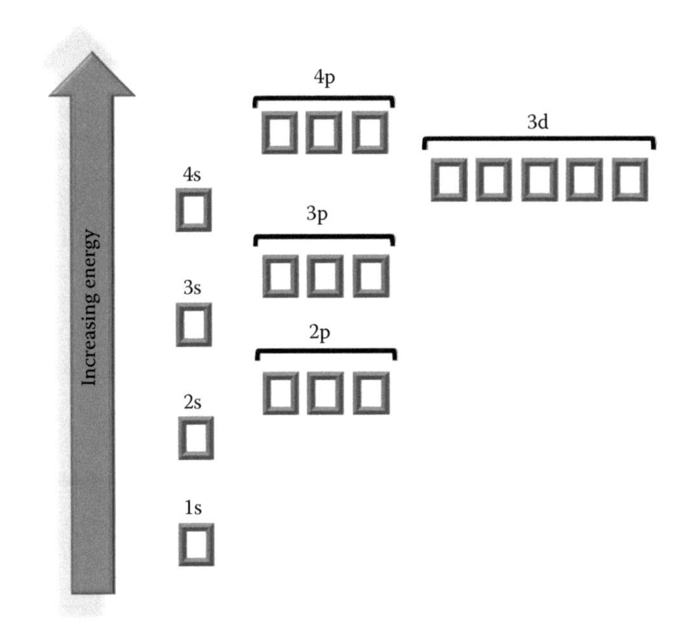

FIGURE 3.7
Arrangement of orbitals in an atom according to increasing energy.

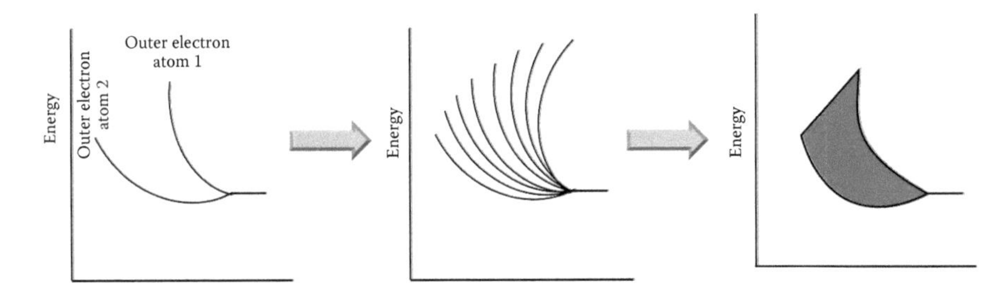

FIGURE 3.8
Splitting of atomic orbitals, forming bands.

3.5.1 Band Diagrams

Orbital arrangement can be visualized using a band diagram (Figure 3.9). The lower-energy band, the valence band, consists of orbitals containing electrons. The higher-energy band consists of unfilled orbitals and is referred to as the conduction band. Depending on the nature of the material, there can be a space between the conduction band and the valence band known as the band gap (Rogers, Adams and Pennathur 2013). When electrons are given sufficient energy they are promoted to the conduction band, and are now able to move through the material (Rogers, Adams and Pennathur 2013).

Nonconducting materials (insulators) have a wide band gap (Figure 3.10a) decreasing the probability of electron transfer from the valence band to the conduction band (Rogers, Adams and Pennathur 2013). Insulators are poor conductors of electricity due to the large band gap (Rogers, Adams and Pennathur 2013). Semiconducting materials have smaller band gaps (Figure 3.10b) (Rogers, Adams and Pennathur 2013). In this case, heat or electricity can facilitate the movement of electrons from the valence band to the conduction band (Rogers, Adams and Pennathur 2013). The smaller

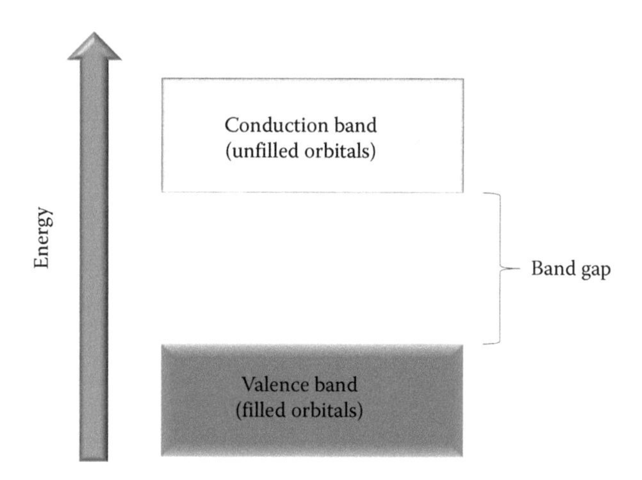

FIGURE 3.9
Band diagram.

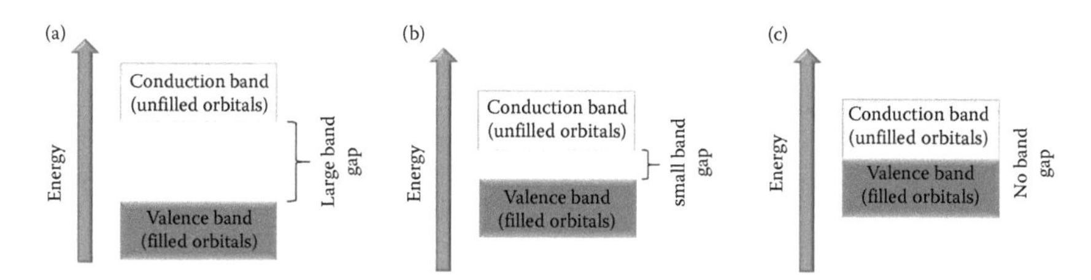

FIGURE 3.10
Band diagrams for insulators (a), semiconductors (b), and metals (c).

band gap of semiconductors increases the likelihood of an electron crossing the gap (Rogers, Adams and Pennathur 2013). In conductors, there is a direct overlap between the valence band and the conduction band (Figure 3.10c). Since the conduction band and the valence band are not separated by a gap; these materials are referred to as zero-band gap materials (Rogers, Adams and Pennathur 2013). Due to the absence of a band gap, these materials are good conductors of heat and electricity (Rogers, Adams and Pennathur 2013). In Chapter 6, band diagrams will be used to explain the unique optical properties of gold nanoparticles and quantum dots.

3.5.2 Fermi Level

Free electrons are present in all materials (Rogers, Adams and Pennathur 2013). There are few free electrons present in insulators, while free electrons are present in abundance in conductors, and semiconductors have a moderate number of free electrons. Although free electrons are delocalized, they must occupy a specific quantized energy state (Rogers, Adams and Pennathur 2013). The Fermi Level marks the probability of finding a free electron in a given energy state (Rogers, Adams and Pennathur 2013). There is a 100% chance of finding a free electron below the Fermi Level (E_f) and no chance of finding a free electron above the Fermi Level (Rogers, Adams and Pennathur 2013). The Fermi Level is located in the band gap of insulators and semiconductors as shown in Figure 3.11 (Rogers, Adams and Pennathur 2013). Only valence electrons in close proximity to the Fermi Level are excited above it. Electrons must be within k_bT of the Fermi Level to be excited above it. The Boltzman constant, k_b, is 1.38×10^{-23} J/K, and T is the temperature in Kelvin (Rogers, Adams and Pennathur 2013). When the average spacing between the energies of free electrons energies and the Fermi Level (ΔE) is larger than k_bT, the electrons remain below E_f, and no conduction occurs (Rogers, Adams and Pennathur 2013).

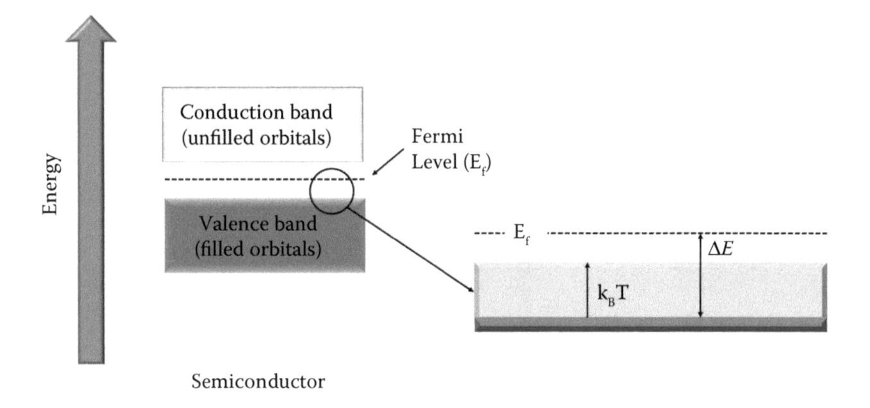

FIGURE 3.11
Location of the Fermi Level in band diagrams.

3.5.3 The Bohr-Exciton Radius

When a material absorbs a photon of sufficient energy, electrons are promoted from the valence band to the conduction band, creating an electron-hole pair (Kittel 2005). This electron-hole pair is called an exciton (Figure 3.12). Photon energy must be greater than the band gap to create excitons. Radiative recombination occurs when the electron drops into the hole in the valence band, emitting a photon in the process (Kittel 2005). When an electron travels to the conduction band, the empty location is referred to as a hole (Rogers, Adams and Pennathur 2013). Electrons are negatively charged and when they leave the valence band the hole left behind assumes a positive charge (Rogers, Adams and Pennathur 2013).

The hole-electron distance is referred to as the Bohr-exciton radius (a_0), as defined by Equation 3.4 (Koole et al. 2014).

$$a_0 = \frac{\mathrm{h}^2\varepsilon}{4\pi^2 e^2}\left(\frac{1}{m_e} + \frac{1}{m_h}\right) \tag{3.4}$$

The terms m_e and m_h are the effective masses of electron and hole, respectively. Further, e is the electron charge and ε is the dielectric constant of the

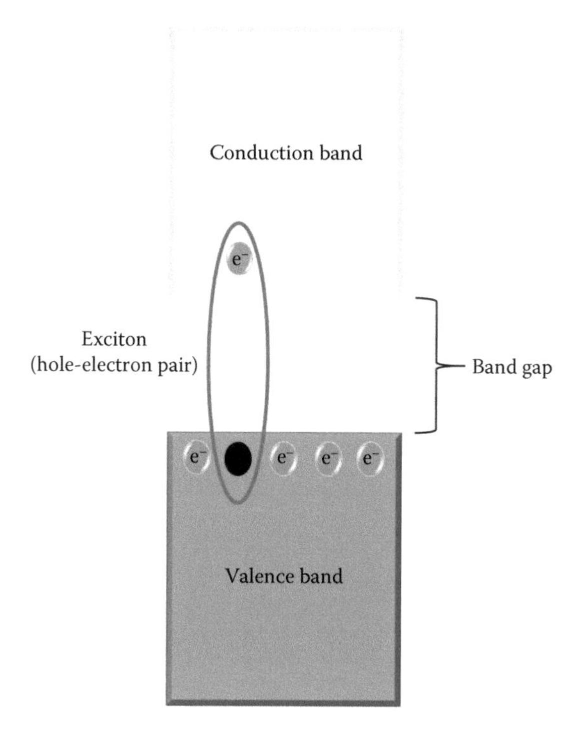

FIGURE 3.12
Hole-electron (exciton) pair.

semiconductor (Koole et al. 2014). Generally, the Bohr-exciton radius is miniscule compared to the dimensions of the solid (Rogers, Adams and Pennathur 2013). The Bohr-exciton radius ranges from 2 to 50 nm depending on the nature of the material. It is interesting to note that the exciton Bohr radius a_0 and the band gap of the semiconductor are correlated, resulting in the fact that materials with wider band gaps possess smaller a_0 (Koole et al. 2014). The Bohr-exciton radius is useful in the determination of the optical properties of quantum dots.

3.6 End-of-Chapter Questions

1. The speed of light is:
   ___ 3.0×10^5 m/s
   ___ 3.0×10^6 m/s
   ___ 3.0×10^7 m/s
   ___ 3.0×10^8 m/s
2. Electromagnetic radiation consists of _____ fields.
   ___ electric
   ___ magnetic
   ___ electric and magnetic
3. The energy of electromagnetic waves can be determined using which of the following equations? Choose all that apply.
   ___ $E = h/\nu$
   ___ $E = \nu/h$
   ___ $E = h\nu$
   ___ $E = h + \nu$
   ___ $E = hc/\lambda$
4. The relationship between the speed of light, frequency, and wavelength is described using:
   ___ $c = \nu/\lambda$
   ___ $c = \lambda/\nu$
   ___ $c = \lambda\nu$
   ___ $c = \lambda + \nu$
5. The double slit experiment provided experimental evidence of the _____ nature of light.
   ___ particle
   ___ wave

6. The photoelectric effect illustrates the _____ nature of light.

    ___ particle

    ___ wave

7. Higher energy bands consisting of unfilled orbitals are referred to as the _____ band.

    ___ conduction

    ___ valence

8. The _____ band is higher energy than the _____ band.

    ___ valence band, conduction band

    ___ conduction band, valence band

9. The band gap:

    ___ is the space between the conduction band and the valence band

    ___ is the energy required to move electrons from the valence band to the conduction band

    ___ can be small or large

    ___ all of the above

10. Which of the following are zero band gap materials?

    ___ conductors (metals)

    ___ semiconductors

    ___ insulators

    ___ none of the above

11. The region in the band diagram that marks the chance of finding a free electron is:

    ___ the band gap

    ___ the Fermi Level

    ___ the conduction band

    ___ the LUMO

12. An exciton is formed when:

    ___ an electron travels from the conduction band to the valence band

    ___ an electron is completely removed from an atom

    ___ an atom gains an electron

    ___ an electron travels from the valence band to the conduction band

13. The hole–electron distance in an exciton is described using:

    ___ the Fermi Level

    ___ the Work Function

    ___ the Bohr-exciton Radius

    ___ tunneling

14. The Bohr-exciton radius ranges from _____ nm.
    ___ 1–10
    ___ 3–5
    ___ 2–50
    ___ 1–100

References

Brown, T. L., H. E. LeMay, Jr., B. E. Bursten, C. J. Murphy, and P. M. Woodward. 2012. *Chemistry—The Central Science*. Boston: Prentice Hall.

Kittel, C. 2005. *Introduction to Solid State Physics*. Hoboken: John Wiley and Sons.

Koole, R., E. Groeneveld, D. Vanmaekelbergh, A. Meijerink, and C. M. Donegá. 2014. "Size effects on semiconductor nanoparticles." In *Nanoparticles—Workhorses of Nanoscience*, edited by C. M. Donegá, 13–51. Berlin: Springer-Verlag.

Rogers, B., J. Adams, and S. Pennathur. 2013. *Nanotechnology The Whole Story*. Boca Raton: CRC Press/Taylor & Francis Group.

Serway, R. A., C. Vuille, and J. S. Faughn. 2006. *College Physics*. Belmont: Brooks/Cole.

4

Allotropic Carbon-Based Nanomaterials

Key Objectives

- Become familiar with the discovery of various carbon-based nanomaterials
- Know the basic structure and properties of carbon-based nanomaterials
- Understand the basic chemical and physical phenomena responsible for the unique mechanical, electrical, and chemical properties of carbon-based nanomaterials
- Know how carbon-based nanomaterials are made
- Understand how CNT electronics function

4.1 Introduction

The word carbon is based on the Latin word *carbo* meaning charcoal (Spyrou and Rudolf 2014). There are many forms of pure carbon with nanoscale dimensions. This chapter will describe the discovery, structure, and unique properties of carbon-based nanomaterials.

4.2 Carbon Allotropes

Before examining nanoscale forms of carbon, it is important to discuss the most common forms of this element. Naturally occurring carbon-based materials exist in the form of allotropes, multiple forms of the same element with different atomic structures (Dume and Tyrell 2012). Two common allotropes of carbon are graphite and diamond (Dume and Tyrell 2012).

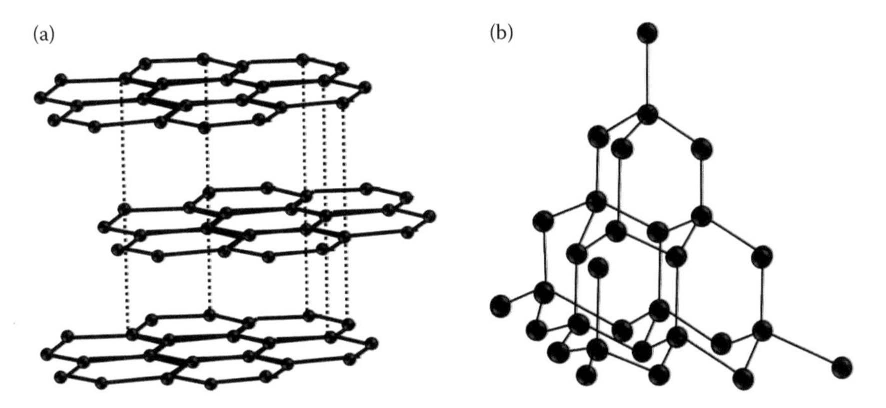

FIGURE 4.1
Carbon allotropes of graphite (a) and diamond (b).

The atomic structure of graphite consists of carbon atoms bonded to three neighboring atoms, as shown in Figure 4.1a. This atomic arrangement forms sheets of hexagonally arranged carbon atoms. These sheets are stacked on top of each other and held together by London dispersion forces. Diamond consists of carbon atoms bonded to four neighboring carbon atoms, resulting in a 3D atomic structure (Figure 4.1b).

4.3 C_{60}

The discovery of C_{60} is a result of the work conducted in 1985 by Robert F. Curl, Harold W. Kroto, and Richard E. Smalley who used lasers to recreate carbon nucleation events similar to those occuring in red giant stars (Delgado et al. 2014). Professor Richard Smalley developed a laser vaporization cluster-beam apparatus (Figure 4.2) enabling the study of stable clusters formed in a plasma. The plasma was produced by a pulsed laser focused on a solid target (Kroto 1992). Upon meeting professor Smalley, professor Kroto saw the cluster beam apparatus and thought replacing silicon with carbon would allow the reproduction of chemical reactions found in space (Kroto 1992). A high-powered laser was used to blast graphite, and the resulting soot was analyzed revealing the presence of a closed cage consisting of 60 carbon atoms (Kroto 1992). Due to the molecule's similarity to the geodesic dome (Figure 4.3), C_{60} molecules were named fullerenes after Buckminster Fuller, the architect who created the structure. Professors Kroto, Smalley, and another Rice University faculty member associated with the project, Bob Curl, won the 1996 Nobel prize in chemistry for the discovery of C_{60} (Talbot 1999).

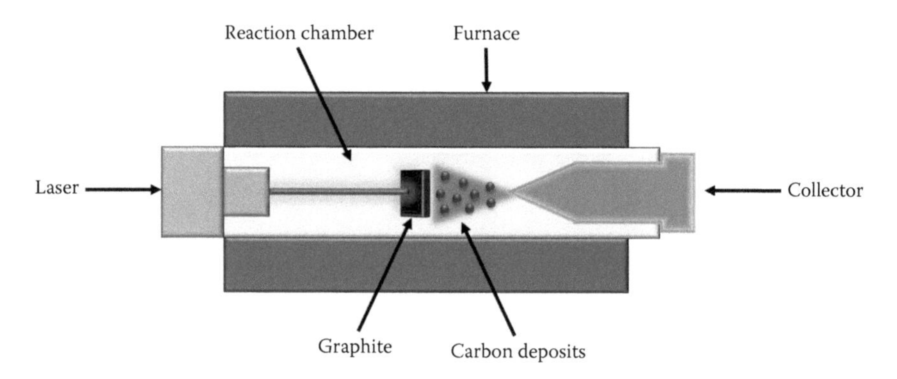

FIGURE 4.2
Laser ablation synthesis of C_{60}.

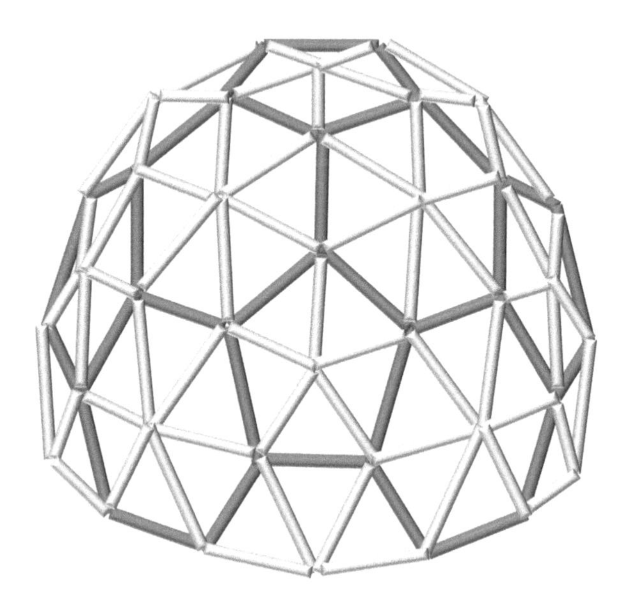

FIGURE 4.3
Geodesic dome.

In 1990, the first successful preparation of macroscopic (gram) quantities of C_{60} was reported by Krätschmer and Huffman (Parker et al. 1991; Meier and Selegue 1993; Hirsch 2010). Krätschmer and coworkers developed the arc-discharge method (Dubrovsky and Bezmelnitsyn 2003) (Figure 4.4). This technique involves producing an arc between two graphite electrodes in a helium atmosphere (Kratschmer et al. 1990). This method utilizes a welding transformer, a glass chamber connected to a vacuum pump, and graphite rods approximately 6 mm in diameter (Pradeep 2007). The graphite rods are brought into close proximity facilitating the formation of an arc. The arc is

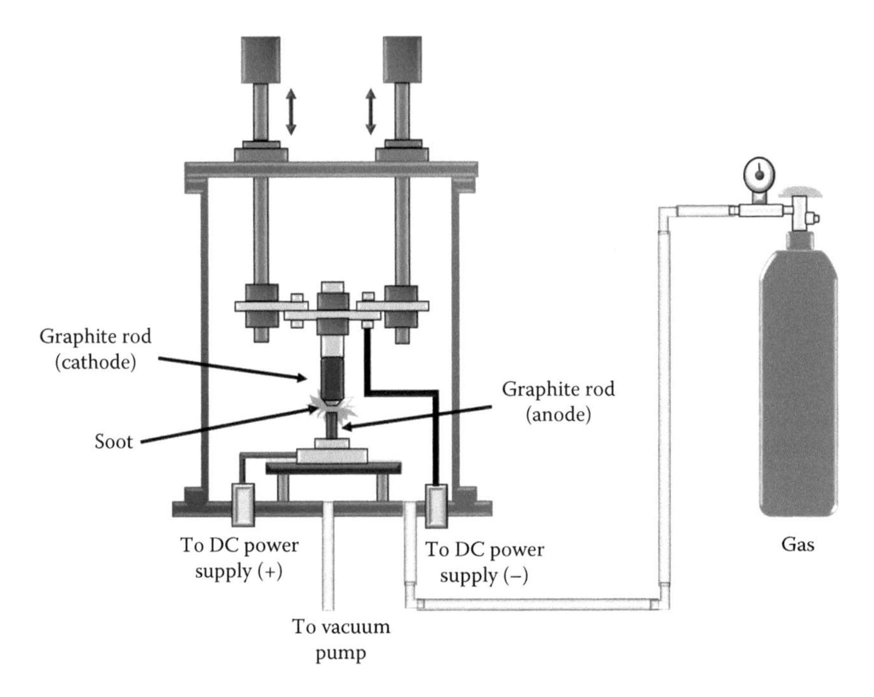

FIGURE 4.4
Kratschmer–Huffman (arc discharge) apparatus.

maintained in a helium or argon atmosphere. It is reported that a voltage of 20 volts at a current of 50–200 amperes was applied across the graphite electrodes to sustain an arc for fullerene production (Pradeep 2007). The soot collected from this method is added to toluene or benzene for 5–6 hours for C_{60} extraction. This final step results in the formation of a dark, reddish-brown solution containing C_{60} molecules (Pradeep 2007).

4.3.1 C_{60} Structure

C_{60} is a spherical carbon molecule with 60 carbon atoms, which surprisingly enough also resembles a soccer ball (Figure 4.5a). C_{60} contains 12 five-membered rings isolated by 20 six-membered rings (Figure 4.5b) (Talbot 1999; Wilson et al. 2002). Each hexagon has three pentagons and three hexagons as its neighbors, whereas pentagons are surrounded by hexagons. This structure provides C_{60} molecules with extraordinary stability (Vintin, Sementsoz and Vagidov 2010). C_{60} molecules have diameters of 0.7 nm and have alternating systems of 60 single and 30 double bonds (Talbot 1999).

4.3.2 C_{60} Properties

Carbon atoms found in C_{60} have four valence electrons, three of which participate in chemical bonding (Delgado et al. 2014). The fourth electron can form a

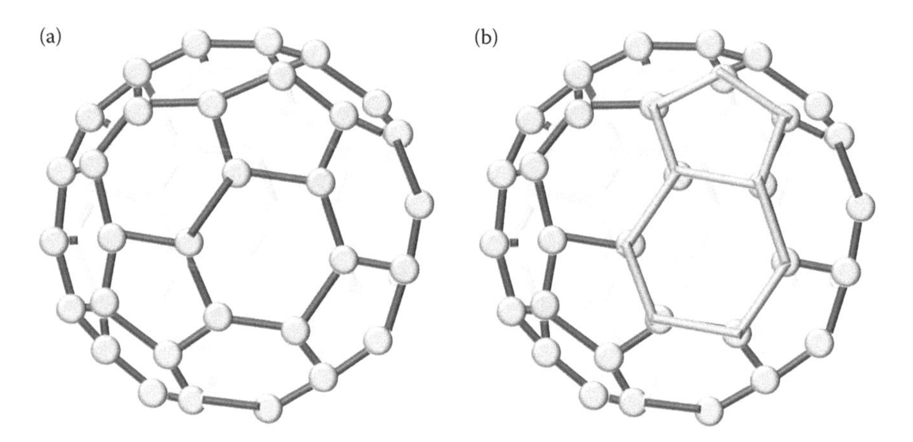

FIGURE 4.5
Structure of C_{60} molecule (a). C_{60} structure indicating positions of pentagons and hexagons (b).

delocalized chemical bond through the formation of π bonds. These π bonds facilitate the delocalization of electrons allowing them to move freely around other carbon rings in the molecule (Delgado et al. 2014). The electrons involved in pi (π) bonding contribute to the conduction of the molecule (Hou et al. 2004). C_{60} exhibits high mechanical strength. This is evident when comparing the bulk modulus of C_{60} (668 GPa) to diamond (442 GPa) and steel (160 GPa) (Rouff and Rouff 1991). C_{60} is described as a homogeneous elastic solid (Rouff and Rouff 1991), which means C_{60} molecules retain their original shape after removal of high pressures applied to the structures. Reports state fullerenes can withstand collisions up to 15,000 mph against stainless steel (Rouff and Rouff 1991).

4.3.3 Endohedral Modification

Due to the cage-like structure of C_{60}, they have the potential for use as molecular storage for a variety of substances (Rogers, Adams and Pennathur 2013). Using C_{60} as a "nanocontainer" is an example of endohedral modification, which involves the incorporation of individual atoms, molecules, or ions within the C_{60} cage, as shown in Figure 4.6 (Gimenez-Lopez et al. 2010). There is great difficulty associated with opening a C_{60} cage for encapsulation of foreign atoms. For this reason, endohedral material must be introduced during formation of the cage (Dunsch and Yang 2006; Thakral and Mehta 2006). Additionally, endohedral C_{60} molecules can be synthesized using implantation, which involves the acceleration of ions to the C_{60} cage (Minezaki et al. 2014). Collisions between the foreign particle and the C_{60} cage allows the cage to absorb enough kinetic energy to open the cage and allow the particle to enter. The particle remains trapped because it does not have sufficient kinetic energy to escape (Minezaki et al. 2014). When a metal is enclosed in the C_{60} cage, the charge transfer induced by the metal within the cage can change the reactivity and properties of the fullerene (Infante, Gagliardi and Scuseria 2008).

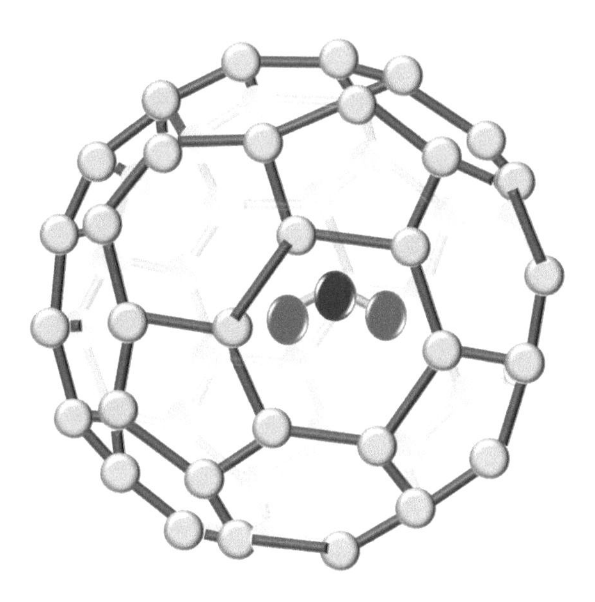

FIGURE 4.6
Endohedral fullerene.

4.3.4 Exohedral Modification

C_{60} molecules can readily bond with atoms of other materials (for example, atoms of hydrogen or fluorine) (Vintin, Sementsoz and Vagidov 2010), which makes it possible to modify the outside of the C_{60} cage (Wilson et al. 2002; Thakral and Mehta 2006). Attachment of atoms or molecules to the surface of the C_{60} cage is referred to as exohedral modification (Figure 4.7) (Hirsch 2010). Generally, when atoms or molecules are added to the outside of the C_{60} cage, these additions occur selectively at the junction of two six-membered rings (Hirsch 2010). One of the many uses of exohedral modification involves improving solubility of C_{60} (Thakral and Mehta 2006). This is important when C_{60} is incorporated in drugs (Hirsch 2010). Bonds in C_{60} can be chemically converted from C = C bonds to C-H bonds, forming fullerene hydrides (Withers, Loufty and Lowe 1997). Fullerene hydrides can lead to improved hydrogen fuel cells, which are safer with increased storage density (Withers, Loufty and Lowe 1997). Additionally, the exohedral attachment of organic molecules, like porphyrins, produces molecules that simulate natural photosynthesis, which can be used in organic solar cells (Hirsch 2010).

4.3.5 Metallofullerenes

Metal-doped C_{60} exhibits superconducting characteristics; this was first reported in 1991 (Takeya et al. 2012). These superconducting structures are

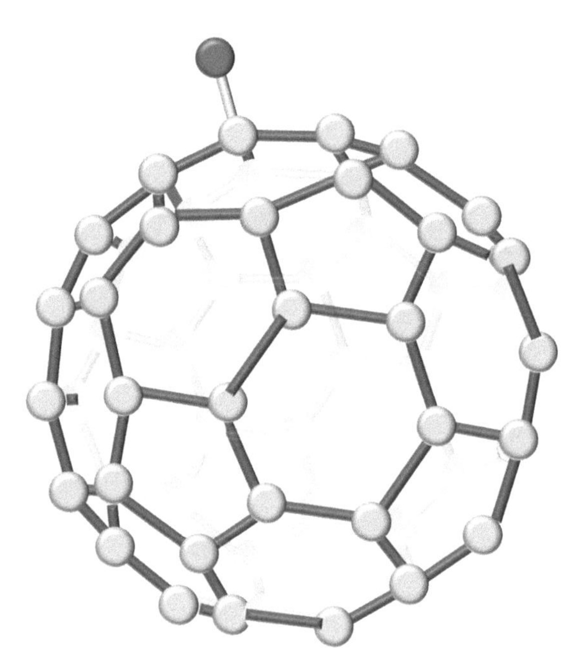

FIGURE 4.7
Exohedral fullerene.

created when C_{60} molecules assemble to form face centered cubic (FCC) crystal structures held together via van der Waals forces (Vintin, Sementsoz and Vagidov 2010; Zadik et al. 2015). The C_{60} crystal is an insulator. However, when atoms of alkali metals are introduced to the crystal structure, C_{60} is converted to a superconductor. Atoms of alkali elements can be easily placed in the empty spaces between C_{60} molecules, as shown in Figure 4.8 (Vintin, Sementsoz and Vagidov 2010). An example of this synthesis occurs when potassium vapor diffuses into the empty spaces between the C_{60} molecules. This results in the formation of K_3C_{60}, a conductive material. At temperatures below 18 K, the electrical resistance of K_3C_{60} reaches zero, and it becomes a super conductor (Vintin, Sementsoz and Vagidov 2010).

4.4 CNTs

Sumio Iijima is credited for the discovery of CNTs (Ajayan and Zhou 2001). This discovery involved the use of transmission electron microscopy to study the surface of graphite electrodes used in an electric arc discharge apparatus. In his analysis, Iijima observed that the sample contained tubules, similar

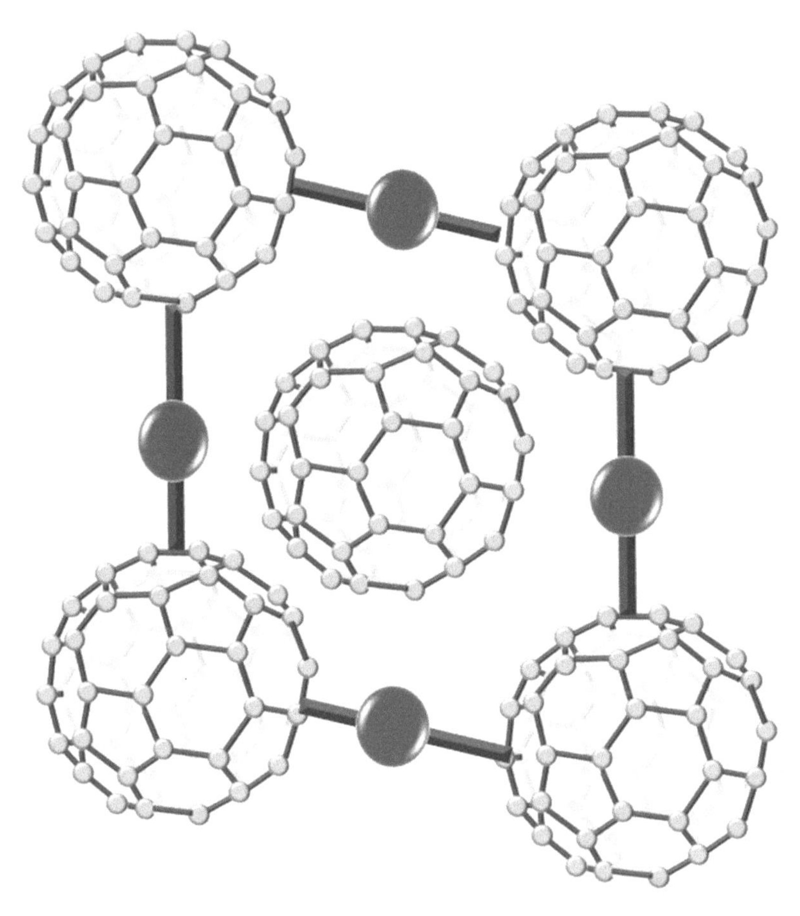

FIGURE 4.8
Crystal structure of a superconducting face centered cubic crystal containing C_{60} molecules.

to those seen in the transmission electron microscope image in Figure 4.9. Cylindrical carbon tubes and fibers were known at the time, but these carbon tubes were different (Ajayan and Zhou 2001). These nanotubes appeared to be capped at the ends with pentagons. Most importantly, Iijima noticed that the arrangement of carbon atoms in the tubes had varying degrees of helicity (Ajayan and Zhou 2001).

4.4.1 Structure

A nanotube consists of one or more seamless cylindrical shells of graphitic sheets. In other words, each shell is made of a hexagonal arrangement of carbon atoms forming a network without any edges (Ebbeson 1996). Nanotubes can possess either metallic or semiconducting characteristics depending on helicity and diameter. The diameter of a nanotube also affects the mechanical

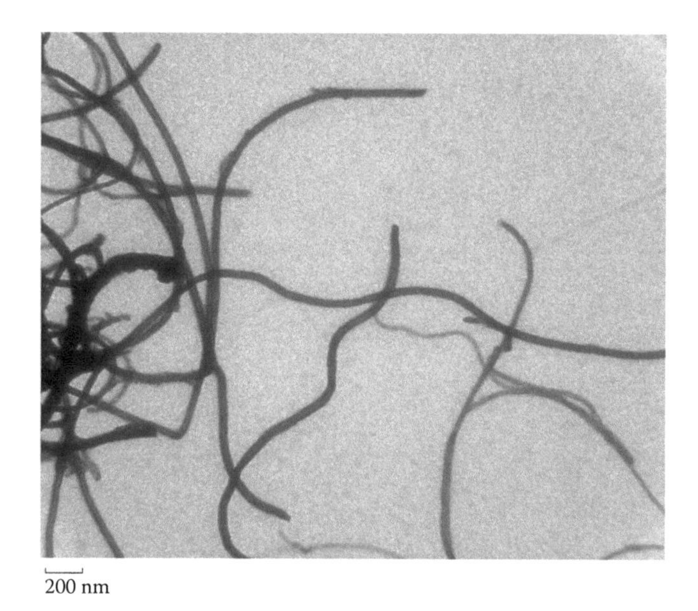

FIGURE 4.9
Transmission electron microscope image of CNTs.

properties of CNTs, and thus can impact applications ranging from scanning probe microscopy tips, electromechanical devices, and structural composites (Cheung et al. 2002). Experimental and theoretical studies suggest that electronic properties, such as the conductance of the tubes, are sensitive to mechanical deformations. CNTs have diameters ranging from 1 to 20 nm (Ebbeson 1996). They are approximately 50,000 times smaller than the width of a human hair (Minezaki et al. 2014), and can be up to 18 centimeters in length (Seetharamappa, Yellappa and D'Souza 2006). CNTs can consist of a single tube, referred to as single-walled CNTs (SWCNTs) (Figure 4.10a), or can contain several concentric tubes called multiwalled CNTs (MWCNTs) (Figure 4.10b) (Kaushik and Majumder 2015). The concentric tubes found in MWCNTs have an interlayer spacing of 3.4 Å (0.34 nm), with an outer diameter ranging from 1 to 50 nm (Seetharamappa, Yellappa and D'Souza 2006). Both SWCNT or MWCNT can easily bundle together to form ropes due to attractive London dispersion forces, the same forces that hold sheets of graphite together (Bellucci and Onorato 2011). CNTs are single sheets of graphite wrapped in cylinder form. The way the sheet wraps is described using a pair of indices (n, m) referred to as the chiral vector, which describes the helicity of a CNT (Seetharamappa, Yellappa and D'Souza 2006). The numbers n, m of a SWCNT define a vector $\mathbf{C} = n\mathbf{a}_1 + m\mathbf{a}_2$, which shows the direction in which a graphene sheet needs to be rolled up to create a SWCNT with a specific helicity (Figure 4.11) (Roch et al. 2014). Based on the chiral vector, three nanotubes variations are possible, armchair, zigzag, and

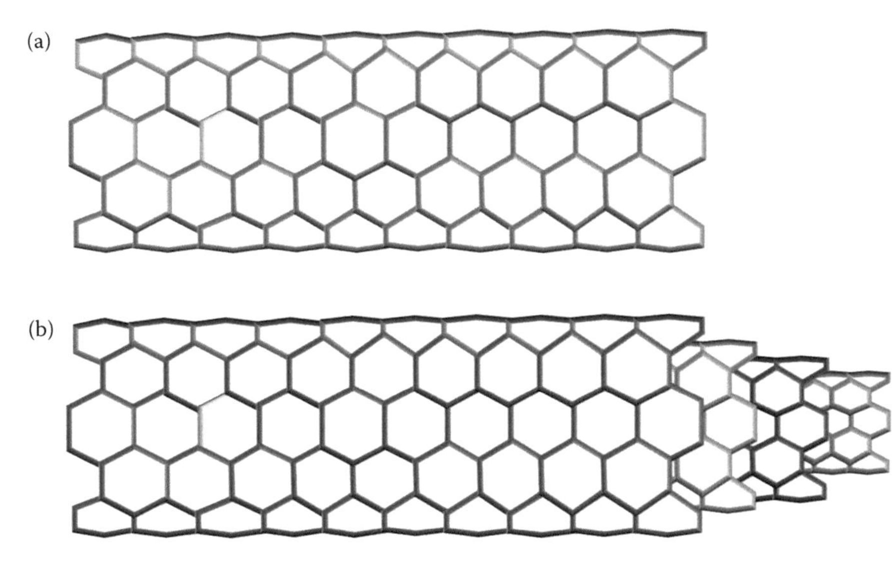

FIGURE 4.10
Single-walled CNT (SWCNT) (a). Multiwalled CNT (MWCNT) (b).

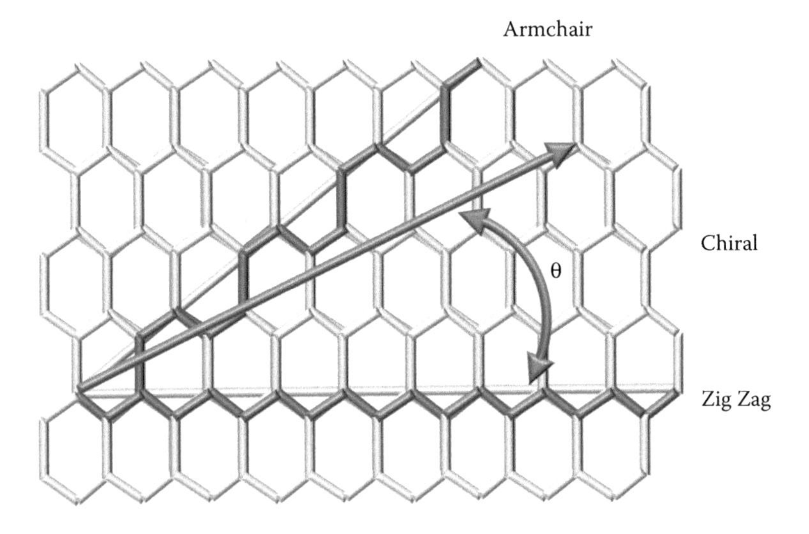

FIGURE 4.11
Indices used to determine the chiral vector (C) and chiral angle (Θ) in a CNT.

chiral (Bellucci and Onorato 2011). These nanotube configurations can also be defined by the chiral angle (θ). Zig-zag nanotubes have a chiral angle of 0°, armchair nanotubes have a chiral angle of 30°, and chiral nanotubes have a value between 0° and 30° (Raifee and Pourazizi 2014). CNT structures are illustrated in Figure 4.12.

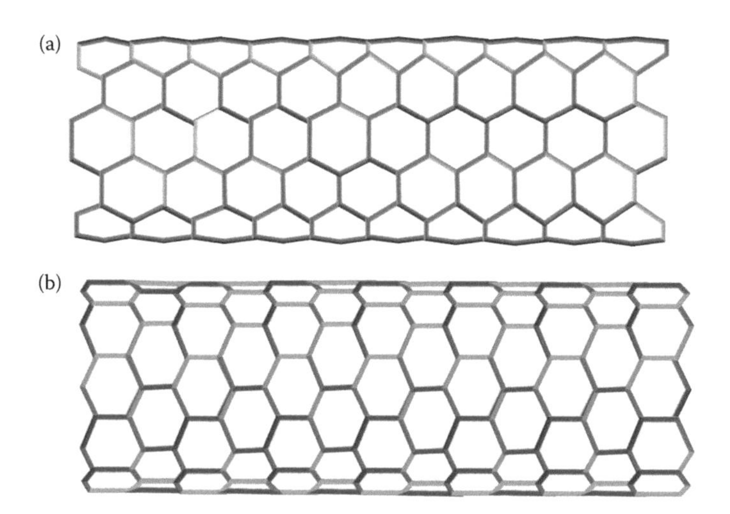

FIGURE 4.12
Armchair (a) and zig zag (b) CNT structures.

4.4.2 Mechanical Properties

CNTs have remarkable mechanical properties. It is reported that CNTs are stronger and stiffer than any substance known (Natsuki 2017). Applying a force on the tip of the nanotube will cause it to bend without breaking; removal of the force will allow the tube to recover its original state (Wilson et al. 2002). Furthermore, under the application of large loads, CNTs bend at extremely large angles (Forro and Schnonenburger 2001). Although rippling and kinks start to appear, the deformations completely disappear once the load is removed (Forro and Schnonenburger 2001). SWCNTs are stronger than steel and are resistant to damage from physical forces, as evidenced by the elastic moduli and tensile strengths listed in Table 4.1 (Wilson et al. 2002; Seetharamappa, Yellappa and D'Souza 2006). For this reason, consumer products such as tennis rackets, bicycle components, golf clubs, and car bumpers use CNTs as strength additives (Rogers, Adams and Pennathur 2013). The high mechanical strength of CNTs is attributed to the strong C-C covalent bonds throughout the structures (Avouris et al. 2003).

TABLE 4.1
Mechanical Properties of Carbon Nanotubes and Other Materials

Material	Modulus of Elasticity (GPa)	Tensile Strength (GPa)
CNT (SWCNT)	1054	150
CNT (MWCNT)	1200	150
Steel	208	0.4
Epoxy	3.5	0.005
Wood	16	0.008

4.4.3 Electrical Properties

In 1991, a research group at the U.S. Naval Research Laboratory submitted a theoretical paper describing the electronic structure of CNTs. The group predicted that a tube consisting of a single sheet of graphite would have a carrier density similar to metals (Hod, Rabani and Baer 2003). To understand the electronic structure of CNTs, consider a single sheet of graphite. Carbon has four valence electrons, three of which are covalently bonded to neighboring atoms. The fourth electron is delocalized and shared by all the atoms, via mobile π bonds (Forro and Schnonenburger 2001). These delocalized π electrons are responsible for the increase of the electrical conductivity of the CNTs. It has also been determined that the conductivity of a SWCNT varies with diameter (Wilson et al. 2002). The movement of electrons through SWCNTs occurs ballistically (Seetharamappa, Yellappa and D'Souza 2006). Ballistic transport refers to the movement of electrons through a material where scattering, by atoms, molecules, or impurities is absent (Bellucci and Onorato 2011). Particles travelling ballistically are characterized by a large mean free path. For instance, the maximum mean free path measured experimentally at room temperature in a CNT was determined to be 1 μm (Lekawa-Raus et al. 2014). By comparison, electrons in copper at room temperature have a mean free path of 40 nm (Lekawa-Raus et al. 2014). Ballistic transport enables nanotubes to carry high currents with little to no heating (Seetharamappa, Yellappa and D'Souza 2006). A metallic CNT can be considered as a highly conductive material (Seetharamappa, Yellappa and D'Souza 2006). The conductivity of metallic CNTs can be more than 50 times higher than copper (Rogers, Adams and Pennathur 2013). It is important to note there are some limitations to electron transport in CNTs (Venema et al. 1999). CNT helicity affects the conductivity of the tube. Armchair CNTs have metallic properties; chiral and zig-zag nanotubes have properties of semiconductors (Kaushik and Majumder 2015).

4.5 CNT Electronics

4.5.1 Transistors

CNTs offer considerable improvements to the field of semiconductors and electronics. Higher storage density, faster processing speeds, and higher energy efficiencies are all possible with CNT-based electronics. To understand the potential advantages of CNT-based electronics, an understanding of the operation of transistors is necessary. Transistors are the most ubiquitous and versatile piece of electronics ever created (Rogers, Adams and Pennathur 2013). A transistor is a three-terminal semiconductor device used to amplify and switch electronic signals and electrical power (Gupta 2013). Currently, integrated circuits (ICs) in personal computers have over a billion transistors, each one turning on

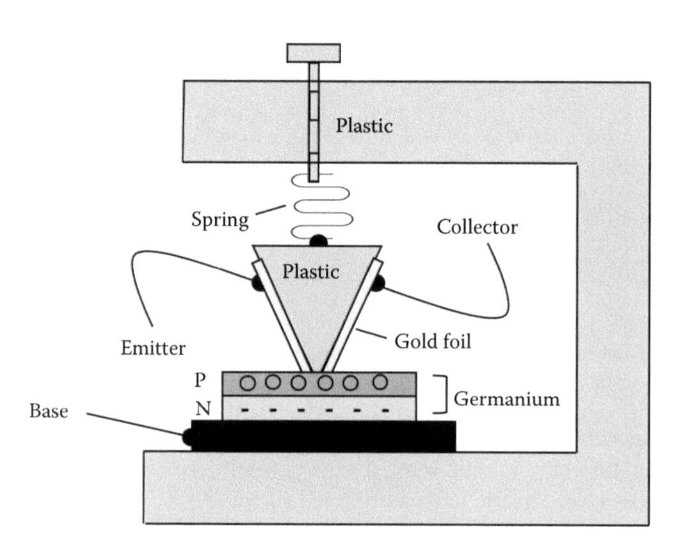

FIGURE 4.13
Diagram of the first point-contact transistor.

and off a billion times every second (Wilson et al. 2002). These on and off states generate the ones and zeros digital computers need for calculations (Kumar and Dubey 2013). Since the invention of the transistor, the semiconductor industry has affected nearly every aspect of our daily life (Chae and Lee 2014). The first transistor was invented in 1947 by William Shockley, John Bardeen, and Walter Brattain (Figure 4.13); they were awarded the Nobel prize in physics in 1956 for their invention (Rogers, Adams and Pennathur 2013). The first transistor was approximately a centimeter high and made of two gold wires separated by 0.02 inches on a germanium crystal (Riordan and Herring 1999; Wilson et al. 2002).

With the invention of transistors in 1947, the field of electronics shifted from vacuum tubes to solid-state devices. As the complexity of circuits increased, engineers started to run into problems (Kahn, Saeed and Ansari 2016). In the late 1960s, Gordon Moore, the cofounder of Intel Corporation, made a prediction that has since become known as Moore's Law (Wilson et al. 2002). Moore predicted that the number of transistors on a chip would double every 18–24 months due to the decrease in feature sizes of electronic components. A consequence of Moore's Law is as transistors get smaller they contain fewer and fewer electrons (Wilson et al. 2002). Eventually a limit of one electron per "bit" will be reached. For this reason, the field of nanoelectronics emerged. This field involves building electronic devices at the atomic level to harness these small-scale "quantum" properties of nature (Wilson et al. 2002).

4.5.2 MOSFETs

Metal oxide semiconductor field effect transistors (MOSFETs) are the most commonly used transistors (Rogers, Adams and Pennathur 2013). MOSFETs,

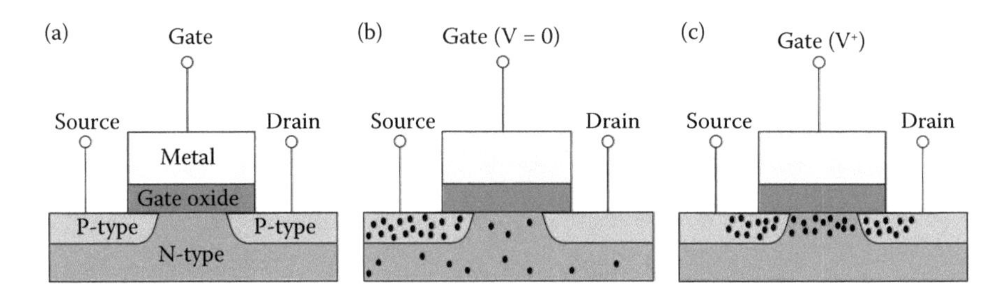

FIGURE 4.14
Diagram of n-channel MOSFET (a). Movement of electrons through a MOSFET in the off-state (b), and on-state (c).

shown in Figure 4.14a, were first developed at Bell Laboratories in the 1960s. In MOSFETs, an electric field on a metal gate electrode is used to control the number of electrons in the active channel of the device. The gate electrode is placed on top of an insulator (silicon dioxide), which is grown on the silicon semiconductor surface (Rogers, Adams and Pennathur 2013). With no voltage on the metal electrode there are no free electrons in the silicon to conduct, thus the transistor is in its "off" state (Figure 4.14b). By applying a positive voltage to the metal gate electrode, electrons are attracted towards the gate from the source and drain electrodes (Figure 4.14c). The insulating layer of silicon dioxide prevents them from reaching the gate so they form a thin, two-dimensional sheet of electrons underneath the oxide layer (Rogers, Adams and Pennathur 2013). These electrons form a conducting channel between source and drain, allowing a current to flow. The transistor now conducts electricity and can be considered in its "on" state. The transistor can be turned between on and off states repeatedly just by changing the applied voltage (Wilson et al. 2002).

4.5.3 CNTFETs

CNT field effect transistors (CNTFETs) have displayed exceptional electrical properties which are superior to the traditional MOSFETs (Stokes and Khondaker 2016). CNT electronics were first demonstrated in 1998 (Kahn, Saeed and Ansari 2016) and are being considered as potential replacements to silicon transistors (Prabhu and Sarwade 2013). CNTFETs are projected to outperform current digital electronic technologies. CNTFET-based devices offer high mobility and high carrier velocity for fast switching (Prabhu and Sarwade 2013). CNTFETs are highly promising candidates to complement silicon-based electronics due to their excellent transport properties (Shulaker et al. 2014). CNTFETs are field-effect transistors utilizing single CNTs or nanotube arrays as the channel material instead of bulk silicon (Figure 4.15a,b). The current flowing in this channel can be switched on or off by applying voltages to a nearby third electrode. It is reported that the gate electrode can change the conductivity of the nanotube channel in an CNTFET by a

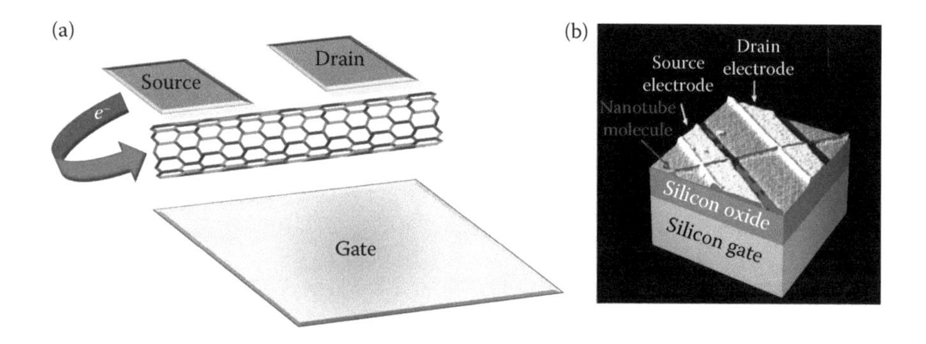

FIGURE 4.15
Diagram of a CNT field effect transistor (CNFET) (a). Atomic force microscope image of a CNT field effect transistor courtesy of Dr. Sander J. Tans, Delft University of Technology.

factor of one million or more (Shulaker et al. 2014). The gate, source, and drain are manufactured using traditional lithography techniques discussed in Chapter 8.

The remarkable electrical properties of SWCNTs stem from the unusual electronic structure of the 2D material graphene. It has a bandgap in most directions but has a vanishing bandgap along specific directions. For this reason, graphene is referred to as a zero-bandgap semiconductor (McEuen and Park 2004). Due to the small size, CNFETs can operate using less power than a silicon-based device (Collins and Avouris 2000). Theorists predict that a truly nanoscale switch could run at clock speeds of one terahertz or more—1000 times faster than processors available today (Collins and Avouris 2000). Due to the one-dimensional transport and long mean free path (in the order of a few hundred nanometers), CNTs can offer scattering-free ballistic transport resulting in low power dissipation. The good thermal conductivity of SWCNTs also brings benefits for reducing power consumption in devices and circuits (Che et al. 2014).

4.6 CNT Synthesis

4.6.1 Arc Discharge Synthesis

Arc discharge synthesis is the simplest method for the creation of CNTs. This method is also referred to as the Kractschmer–Huffman method, and is also used to synthesize C_{60}. Arc discharge synthesis involves positioning two carbon rods, end to end, with a separation of approximately 1 mm (Kaushik and Majumder 2015). The carbon rods are housed in a glass enclosure, at low pressure and filled with an inert gas. A direct current of 50–100 A is used to produce a high temperature discharge between the two carbon rods (Kaushik and Majumder 2015). Nanotube yield and quality is dependent on the conditions of the arc (Hod, Rabani and Baer 2003). The most important parameters controlling nanotube quality are gas pressure and stability of the arc plasma. Arc discharge synthesis produces the best quality nanotubes when a current of about 50 A is applied between two carbon electrodes in a helium atmosphere. The deposit on the cathode contains the CNTs. If metals such as Co or Ni are added to the anode, SWCNTs are produced in the arc discharge synthesis (Hod, Rabani and Baer 2003).

4.6.2 Laser Ablation Synthesis

Laser ablation, a technique used to synthesize C_{60}, is also used to synthesize CNTs (Kaushik and Majumder 2015). Intense laser pulses are used to ablate a carbon target. Using a laser to ablate graphite in an atmosphere containing an inert gas and a catalyst produces CNTs assembled to form ropes with diameters between 5 to 20 nm diameter and tens to hundreds of micrometers long (Hod, Rabani and Baer 2003).

4.6.3 Chemical Vapor Deposition

Carbon fibers and filaments are also produced using chemical vapor deposition (CVD). This is a nanofabrication technique which will be covered in greater detail in Chapter 8. CNT synthesis using CVD involves the reaction of a carbon-containing gas (such as ethane, acetylene, ethylene, ethanol, etc.) with a nanoparticle catalyst (usually, cobalt, nickel, or iron) (Hod, Rabani and Baer 2003). Decomposition of a hydrogen/methane (H_2/CH_4) mixture over cobalt, nickel, or iron nanoparticles at 1000°C is used to obtain high yields of SWNTs (Kaushik and Majumder 2015). The process of forming nanotubes using CVD is extensively used for two main reasons. First, nanotubes are obtained in large quantity, and at much lower temperatures than arc discharge or laser ablation (Forro and Schnonenburger 2001). Second, the catalyst (for example iron, cobalt, or nickel) can be selectively positioned on the substrate prior to growth; this allows directed-growth of CNTs (Forro and Schnonenburger 2001).

4.7 CNTs in Medicine

CNTs can transport drug molecules, proteins, and nucleotides because molecules can be covalently attached to the surface of the tubes (Jain et al. 2010). Additionally, the hollow structure of CNTs allows encapsulation of molecules but as yet there are very few examples of this for drug delivery

(Jain et al. 2010). In addition, their size and shape allow them to enter living cells without causing cell death or obvious damage (Jain et al. 2010). Several drugs have been successfully delivered due to covalent or noncovalent functionalization, which prevents aggregation and increases solubility (Jain et al. 2010). Charged biomolecules can be detected by nanowire FETs and CNTFETs (Leyden et al. 2010). In the class of charge-sensitive biosensors, semiconducting CNTs are extremely promising. CNTs are extremely stable in biological environments. The conductance of semiconducting CNT can be tuned either by ions in the liquid environment or by charged proteins binding to the CNT surface (Leyden et al. 2010). Charge carriers in CNTs are capacitively coupled to charges in the environment, and as a result, the number of free carriers is extremely sensitive to changes in the environment. Protein binding can have additional effects on the conductance of a CNT FET device (Leyden et al. 2010).

4.8 Graphene

Graphene is a single atomic layer of graphite (Obeng and Sirinivasan 2011). Graphite consists parallel sheets of carbon held together by weak van der Waals forces (Dume and Tyrell 2012; Spyrou and Rudolf 2014). This explains why graphite is an effective writing medium (Neto et al. 2009). When writing with a number two pencil, graphene stacks are deposited on the paper, and quite possibly as individual graphene layers (Neto, Guinea and Peres 2006). Since its discovery, scientists have learned that graphene has very unusual and interesting electrical and mechanical properties (Kumar et al. 2013).

4.8.1 Graphene Discovery

As early as 1947, it was believed graphene had extraordinary electronic properties, but the problem was isolating a single sheet of graphite (Obeng and Sirinivasan 2011). For many years, graphene was a material which only existed in theory and was believed to have an unstable nature (Obeng and Sirinivasan 2011). In 2004, Andre Geim and Konstantin Novoselov from Manchester University were the first to successfully isolate a single atomic layer of graphene (Obeng and Sirinivasan 2011). They isolated graphene layers using micromechanical cleavage (Spyrou and Rudolf 2014). This method involved delicately cleaving a sample of graphite with Scotch tape, which led to the production of something long considered impossible, a sheet of crystalline carbon just one atom thick (Neto et al. 2009). This method allows easy production of graphene sheets up to 100 μm in size (Kanelson 2007). Amazingly, Geim and Novoselov visually observed graphene sheets produced by the micromechanical cleavage technique, even though the optical absorbance of graphene is very low (Allen, Tung and Kaner 2010). This was accomplished by depositing the graphene flakes on silicon dioxide (SiO_2). This produced an interference effect that enhanced the optical contrast under white-light illumination (Allen, Tung and Kaner 2010). Viewing a single sheet of graphene with an optical microscope, even though the material is one atom thick, is possible because graphene contains an extensive amount of π bonds, which facilitates sufficient light absorption when deposited on a layer of silicon dioxide (SiO_2) (Spyrou and Rudolf 2014). Geim and Novoselov later confirmed that the graphene layers they produced were only a few angstroms thick using an atomic force microscope (Obeng and Sirinivasan 2011). They were awarded the 2010 Nobel prize in physics for their discovery (Obeng and Sirinivasan 2011).

4.8.2 Structure

Graphene is a sheet of crystalline carbon, with hexagonally arranged atoms, just one atom thick (Figure 4.16) (Vintin, Sementsoz and Vagidov 2010). It is important to note that it takes at least 10 layers before a sample becomes bulk graphite. It is also noteworthy to mention that there is a 3.35 Å spacing between the stacked sheets (Kumar et al. 2013).

Graphene sheets are much stronger than steel, which is apparent when comparing tensile strength values. It is reported graphene has a tensile strength of 130 GPa (Kumar et al. 2013), whereas the reported tensile strength for steel is 0.400 GPa (Rogers, Adams and Pennathur 2013).

FIGURE 4.16
Two-dimensional hexagonal lattice of graphene.

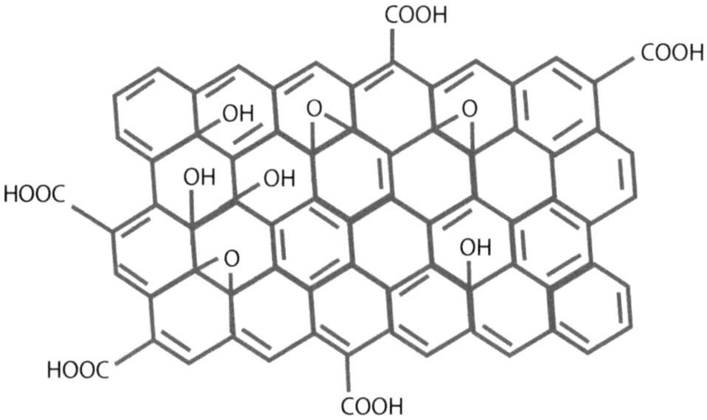

FIGURE 4.17
Structure of graphene oxide.

4.8.3 Graphene Synthesis

Chemical methods are required to synthesize large area graphene films (Kumar et al. 2013). One synthesis technique involves graphite oxide. Graphite oxide (GO) (Figure 4.17) is commonly formed by exposing graphite to strong acids and oxidizing reagents, resulting in the attachment of oxygen groups on the surfaces and at the edges of the graphite layers (Spyrou and Rudolf 2014). GO is then exfoliated in water due to its hydrophilic character. Water molecules can penetrate between the GO layers. Once dispersed, these monolayers of GO act as precursors to produce graphene by the removal of the oxygen groups using a thermal annealing process at temperatures >1000°C (Spyrou and Rudolf 2014).

Another interesting method for the preparation of graphene involves unzipping of CNTs. The resulting structures are referred to as graphene nanoribbons. Unzipping is accomplished by cutting the CNTs along their axis by plasma etching or by chemical means (Spyrou and Rudolf 2014). The most promising, inexpensive approach for deposition of reasonably high-quality graphene is chemical vapor deposition onto metal substrates such Ni, Pd, Ru, and Cu (Kumar et al. 2013). The advantage of this technique is it can be easily extended to large areas by just increasing the metal substrate size (Obeng and Sirinivasan 2011).

4.8.4 Electrical

In 1946, P. R. Wallace was the first to suggest graphene would demonstrate unusual semimetallic behavior (Neto, Guinea and Peres 2006). It is now known that electrons in graphene have a high mobility, 100 times larger than crystalline silicon, the most widely used material in nanoelectronics (Vintin, Sementsoz and Vagidov 2010). This property makes it possible to develop high frequency transistors, devices operating faster than current silicon-based transistors (Dume and Tyrell 2012; Allen, Tung and Kaner 2010). Long-range π bonding throughout graphene results in extraordinary

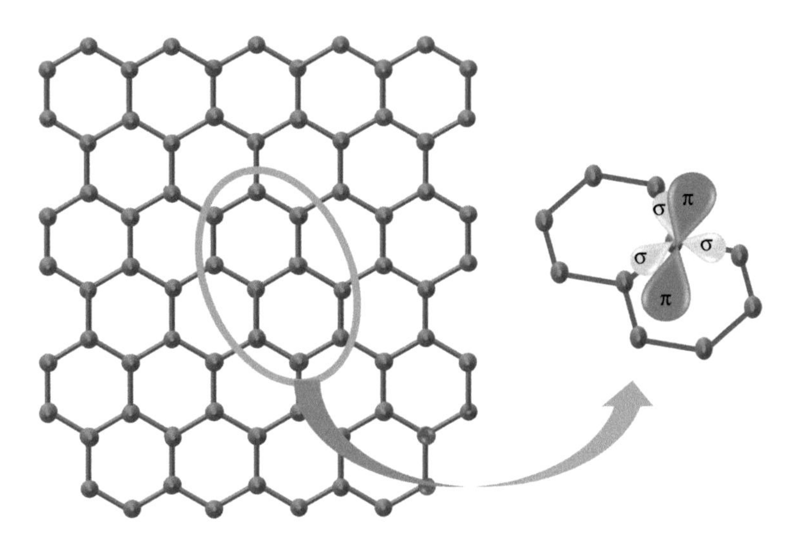

FIGURE 4.18
Arrangement of σ and π bonds in a graphene sheet.

thermal, mechanical, and electrical properties (Allen, Tung and Kaner 2010) (Figure 4.18). As stated in Chapter 2, π bonds are formed from the sideways overlap of p orbitals. The carbon atoms in the graphene layer form three σ (sigma) bonds with neighboring carbon atoms while the remaining p_z orbitals overlap to form a band of filled π bonds, which are responsible for the high *in-plane* conductivity (Spyrou and Rudolf 2014). It is also interesting to point out that the electronic properties of graphene can also be changed by reducing the dimensions of the graphene layers (Spyrou and Rudolf 2014).

4.9 End-of-Chapter Questions

1. C_{60} molecules can be synthesized using:
   ___ laser ablation only
   ___ synthesis in the solution phase only
   ___ arc discharge (Kratchsmer–Huffman) method only
   ___ laser ablation and arc discharge method
2. C_{60} is a caged, spherical structure that contains _____ pentagons and _____ hexagons.
   ___ 10, 30
   ___ 12, 20

___ 30, 10

___ 20, 12

3. The diameters of C_{60} molecules is approximately _____ nm.

___ 10.0

___ 5.3

___ 2.6

___ 0.7

4. Encapuslating atoms, molecules, or ions in C_{60} cages is an example of an _____ modification.

___ exohedral

___ endohedral

5. Attaching atoms or molecules to C_{60} cages is an example of an _____ modification.

___ exohedral

___ endohedral

6. CNT diameters can range from _____ to _____ nm.

___ 2–5

___ 1–2

___ 3–6

___ 1–20

7. A CNT with a chiral angle (η) of 0° has a/an _____ structure.

___ armchair

___ zig-zag

___ chiral

8. A CNT with a chiral angle (η) of 30° has a/an _____ structure.

___ armchair

___ zig-zag

___ chiral

9. A CNT with a chiral angle (η) between 0° and 30° has a _____ structure.

___ armchair

___ zig-zag

___ chiral

10. The conductivity of CNTs can be attributed to:

___ strong covalent bonds between carbon atoms

___ the mechanical strength of CNTs

___ mobile π bonds

___ none of the above

11. The property of CNTs that describes the movement of electrons through the nanotube where scattering by atoms, movements, or impurities is absent is:

___ ballistic electron transport

___ tensile strength

___ bulk modulus

___ π bonding

12. The conductivity of a CNT can be affected by:

___ tensile strength only

___ helicity (structure) only

___ diameter only

___ helicity and diameter

13. Which of the following nanotube structures has the most metallic character?

___ armchair

___ zig-zag

___ chiral

14. In CNT field effect transistors (CNFETs), which of the following terminals can control the flow of current in a CNT?

___ source

___ gate

___ drain

15. Nanotube quality in the arc discharge synthesis is controlled using:

___ gas pressure only

___ plasma stability only

___ graphite rod diameter only

___ gas pressure and plasma stability

16. Single-walled CNTs are synthesized during the arc discharge method when metals are added to the _____.

___ anode

___ cathode

17. The first synthesis of graphene involved the use of:

___ laser ablation

___ micromechanical clevage

___ chemical vapor deposition

___ arc discharge synthesis

18. It is reported that graphene has a modulus (ability to stretch) of _____ gigapascals (GPa), and the modulus of steel is _____ gigapascals (GPa).
    ___ 0.4, 130
    ___ 108, 14
    ___ 130, 0.4
    ___ 208, 1100
19. Synthesis of large area graphene films can involve the use of _____.
    ___ CNTs
    ___ C_{60}
    ___ graphite oxide
    ___ graphene flakes
20. Graphite oxide is converted to graphene using temperatures greater than _____.
    ___ 10°C
    ___ 100°C
    ___ 1000°C
    ___ 10,000°C
21. Graphene has an electron motility that is _____ times larger than silicon.
    ___ 5
    ___ 20
    ___ 100
    ___ 1000
22. The extensive network of _____ bonds found in graphene is responsible for the high electrical conductivity of graphene.
    ___ sigma
    ___ pi

References

Ajayan, P. M., and O. Z. Zhou. 2001. "Applications of carbon nanotubes." *Appl. Phys.* 80: 391–425.

Allen, M. J., V. C. Tung, and R. B. Kaner. 2010. "Honeycomb carbon: A review of graphene." *Chem. Rev.* 110: 132–145.

Avouris, P., J. Apenzeller, R. Martel, and S. J. Wind. 2003. "Carbon nanotube electronics." *Proceedings of the IEEE* 91: 1772–1784.

Bellucci, S., and P. Onorato. 2011. "Transport properties in carbon nanotubes." In *Physical Properties of Ceramic Nanostructures*, 45–109. Berlin: Springer-Verlag.

Chae, S. H., and Y. H. Lee. 2014. "Carbon nanotubes and graphene towards soft electronics." *Nano Convergence* 1: 1–26.

Che, Y., H. Chen, H. Gui, B. Liu, and C. Zhou. 2014. "Review of carbon nanotube nanotube nanoelectronics and microelectronics." *Semicond. Sci. Technol.* 29: 1–17.

Cheung, C. L., A. Kurtz, H. Park, and C. M. Lieber. 2002. "Diameter-controlled synthesis of carbon nanotubes." *J. Phys. Chem. B.* 106: 2429–2433.

Collins, P. G., and P. Avouris. 2000. "Nanotubes for electronics." *Scientific American* 283: 62–69.

Delgado, J. L., F. Filippone, F. Giacalone, M. A. Herranz, B. Illescas, E. M. Perez, and N. Martin. 2014. "Buckyballs." In *Polyarenes II*, edited by J. S. Siegel and Y. T. Wu, 1–64. Switzerland: Springer International Publishing.

Dubrovsky, R., and V. Bezmelnitsyn. 2003. "Enhanced approach to synthesize carbon allotropes by arc plasma." *Rev. Adv. Mater. Sci.* 5: 420–424.

Dume, B., and J. Tyrell. 2012. "20 things you can do with graphene." *Phys. World: Reaping the Benefits of Nanomaterials* June: 11–14.

Dunsch, L., and S. Yang. 2006. "The recent state of endohedral fullerene research." *Electrochem. Soc. Interface* 15: 34–39.

Ebbeson, T. W. 1996. "Carbon nanotubes." *Phys. Today* 49: 26–32.

Forro, L., and C. Schnonenburger. 2001. "Carbon nanotubes: Materials for the future." *Europhys. News* 32: 86–90.

Gimenez-Lopez, M. C., J. A. Gardener, A. Q. Shaw, A. Iwasiewicz-Wabnig, K. Porfyrakis, C. Balmer, G. Dantelle et al. 2010. "Endohedral metallofullerenes in self-assembled monolayers." *Phys. Chem. Chem. Phys.* 12: 123–131.

Gupta, M. 2013. "A study of single electron transistor (SET)." *Int. J. Sci. Res.* 5: 474–479.

Hirsch, A. 2010. "The era of carbon allotropes." *Nat. Mater.* 9: 868–871.

Hod, O., E. Rabani, and R. Baer. 2003. "Carbon nanotube closed-ring structures." *Phys. Rev. B.* 67: 195408–1–195408–7.

Hou, J. G., A. D. Zhao, T. Huang, and L. Shan. 2004. "C60 Based Materials." In *Encyclopedia of Nanoscience and Nanotechnology*, 409–474. Valencia: American Scientific Publishers.

Infante, I., L. Gagliardi, and G. E. Scuseria. 2008. "Is fullerene C60 large enough to host a multiply bonded dimetal?" *J. Am. Chem. Soc.* 130: 7459–7465.

Jain, N., R. Jain, N. Thakur, B. P. Gupta, D. K. Jain, J. Banveer, and S. Jain. 2010. "Nanotechnology: A safe and effective drug delivery system." *Asian J. Pharm. Clincal Res.* 3: 159–165.

Kahn, A. M., S. H. Saeed, and M. S. Ansari. 2016. "A review on carbon nanotubes field effect transistors in different era with related applications." *Int. J. Innov. Res. Electr. Electron. Instrum. Control Eng.* 4: 38–42.

Kanelson, M. I. 2007. "Carbon in two dimensions." *Mater. Today* 10: 20–27.

Kaushik, B. K., and M. K. Majumder. 2015. "Carbon nanotubes: Properties and applications." In *Carbon Nanotube Based VLSI Interconnects: Analysis and Design*, 17–37. Springer, India.

Kratschmer, W., L. D. Lamb, K. Fostiropoulos, and D. R. Huffman. 1990. "Solid C60: A new form of carbon." *Nature* 347: 354–358.

Kroto, H. W. 1992. "Buckminsterfullerene, The celestial sphere that Fell to earth." *Angew. Chem.* 31: 111–129.

Kumar, A., and D. Dubey. 2013. "Single electron transistor: Applications and limitations." *Adv. Electr. Electric Eng.* 3: 57–62.

Kumar, P., A. K. Singh, S. Hussain, K. N. Hui, K. S. Hui, J. Eom, J. Jung, and J. Singh. 2013. Graphene: Synthesis, properties, and applications in transparent electronic devices. *Rev. Adv. Sci. Eng.* 2: 1–21.

Lekawa-Raus, A., J. Patmore, L. Kurzepa, J. Bulmer, and K. Kozial. 2014. "Electrical properties of carbon nanotube based fibers and their future use in electrical wiring." *Adv. Funct. Mater.* 24: 3661–3682.

Leyden, M. R., C. Schuman, T. Sharf, J. Kevek, V. T. Remcho, and E. D. Minot. 2010. "Fabrication and characterization of carbon nanotube field-effect transistor biosensors." In *Organic Semiconductors in Sensors and Bioelectronics III*, edited by R. Shinar and I. Kymissis. San Diego: Proc. of SPIE 7779. 1–12.

McEuen, P. L., and J. Y. Park. 2004. "Electron transport in single-walled carbon nanotubes." *MRS Bulletin* 29: 272–275.

Meier, M. S., and J. P. Selegue. 1993. "Efficient separation of fullerenes on preparative gel permeation chromatography columns." *Energeia* 4: 1–6.

Minezaki, H., S. Ishihara, T. Uchida, M. Muramatsu, R. Racz, T. Asaji, A. Kitgawa, Y. Kato, S. Biri, and Y. Yoshida. 2014. "Synthesis of endohedral iron-fullerenes by ion-implantation." *Rev. Sci. Instrum.* 85: 02A945.

Natsuki, T. 2017. "Carbon nanotube based nanomechanical sensor: Theoretical analysis of mechanical and vibrational properties." *Electronics* 6: 1–20.

Neto, A. C., F. Guinea, and N. M. Peres. 2006. "Drawing conclusions from graphene." *Phys. World* 19: 1–5.

Neto, A. H. C., F. Guinea, N. M. R. Peres, K. S. Novoselov, and A. K. Geim. 2009. "The electronic properties of graphene." *Rev. Mod. Phys.* 81: 109–162.

Obeng, Y., and P. Sirinivasan. 2011. "Graphene: Is it the future for semiconductors? An overview of the material, devices, and applications." *Electrochem. Soc. Interface* 20: 47–52.

Parker, D. H., P. Wurz, K. Chatterjee, K. R. Lykke, J. E. Hunt, M. J. Pellin, J. C. Hemminger, D. M. Gruen, and L. M. Stock. 1991. "High-yield synthesis, separation, and mass spectrometric characterization of fullerenes C60 to C266." *J. Am. Chem. Soc.* 113: 1499–1503.

Prabhu, S., and N. Sarwade. 2013. "Application of CNTFET as logic gates and its implementation using HSPICE." *Int. J. Modern Eng. Res.* 3: 3646–3648.

Pradeep, T. 2007. *Nano: The Essentials Understanding Nanoscience and Nanotechnology.* New Delhi: Tata McGraw-Hill.

Raifee, R., and R. Pourazizi. 2014. "Evaluating the influence of defects on young's modulus of carbon nanotubes using stochastic modeling." *Mat. Res.* 17: 758–766.

Riordan, M. L., and C. Herring. 1999. "The invention of the transistor." *Rev. Mod. Phys.* 71: S336–S345.

Roch, A., L. Stepien, T. Roch, I. Dani, C. Leyens, O. Jost, and A. Leson. 2014. "Optical absorption spectroscopy and properties of single-walled carbon nanotubes at high temperatures." *Synth. Met.* 197: 182–187.

Rogers, B., J. Adams, and S. Pennathur. 2013. *Nanotechnology—The Whole Story.* Boca Raton: CRC Press Taylor & Francis Group.

Rouff, R. S., and A. L. Rouff. 1991. "The bulk modulus of C60 molecules and crystals: A molecular mecahnics approach." *Appl. Phys. Lett.* 59: 1553–1555.

Seetharamappa, J., S. Yellappa, and F. D'Souza. 2006. "Carbon nanotubes: Next generation of electronic materials." *Electrochem. Soc. Interface* 15: 23–26.

Shulaker, M. M., J. V. Rethy, G. Hills, H. Wei, H. Y. Chen, G. Gielen, H. S. P. Wong, and S. Mitra. 2014. "Sensor-to-digital interface built entirely with carbon nanotube FETs." *IEEE J. Solid-State Circuits* 49: 190–201.

Spyrou, K., and P. Rudolf. 2014. "An Introduction to Graphene." In *Functionalization of Graphene*, 1–20. Weinheim: Wiley-VCH.

Stokes, P., and S. I. Khondaker. 2016. "Local-gated single-walled carbon nanotubes field effect transistors assembled by AC dielectrophoresis." *Nanotechnology* 19: 175202.

Takeya, H., K. Miyazaya, R. Kato, T. Wakahara, T. Ozaki, H. Okazaki, T. Yamaguchi, and Y. Takano. 2012. "Superconducting fullerene nanowhiskers." *Molecules* 17: 4851–4859.

Talbot, C. 1999. "Fullerene and nanotube chemistry: An update." *Sch. Sci. Rev.* 81: 37–48.

Thakral, S., and R. M. Mehta. 2006. "Fullerenes: An introduction and overview of their biological properties." *Indian J. Pharm. Sci.* 68: 13–19.

Venema, L. C., J. W. G. Wildoer, J. W. Jassen, S. J. Tans, H. I. J. T. Tuinstra, L. P. Kouwenhoven, and C. Dekker. 1999. "Imaging electron wave functions of quantized energy levels in carbon nanotubes." *Science* 283: 52–55.

Vintin, V. V., D. I. Sementsoz, and N. Z. Vagidov. 2010. *Quantum Mechanics for Nanostructures*. Cambridge: Cambridge University Press.

Wilson, M., K. Kannangara, G. Smith, M. Simmons, and B. Raguse. 2002. *Nanotechnology—Basic Science and Emerging Applications*. Boca Raton: Chapman and Hall CRC.

Withers, J. C., R. O. Loufty, and T. P. Lowe. 1997. "Fullerene commercial vision." *Fullerene Sci. Tech.* 5: 1–31.

Zadik, R. H., Y. Takabayashi, G. Klupp, R. H. Colman, A. Y. Ganin, A. Potocnik, P. Jeglic et al. 2015. "Optimized unconventional superconductivity in a molecular jahn-teller metal." *Condens. Matter Phys.* 1: 1–9.

5

Molecule-Based Nanotechnology

Key Objectives

- To understand how DNA is used to create arbitrary nanostructures
- To become familiar with the structure, fabrication, and use of dendrimers
- To gain familiarity with the applications of dendrimers in the medical field
- To understand how alkanethiol molecules are used to create nanoscale molecular films
- To understand the operation of molecule-based electronics
- To become familiar with various molecule-based drug delivery systems

5.1 Introduction

The natural world has used molecular materials for the assembly of structures for billions of years (Michelotti et al. 2012). Scientists' study and understanding of naturally occurring nanostructures has resulted in the development of methodologies involving the use of molecules as building blocks for the assembly of nanostructures (Michelotti et al. 2012). This chapter describes the synthesis, characteristics, and applications of various molecule-based nanomaterials.

5.2 DNA Nanotechnology

5.2.1 DNA Components

DNA is a 2-nm diameter structure containing genetic information for living organisms (Rogers, Adams and Pennathur 2013). Deoxyribonucleic acid (DNA)

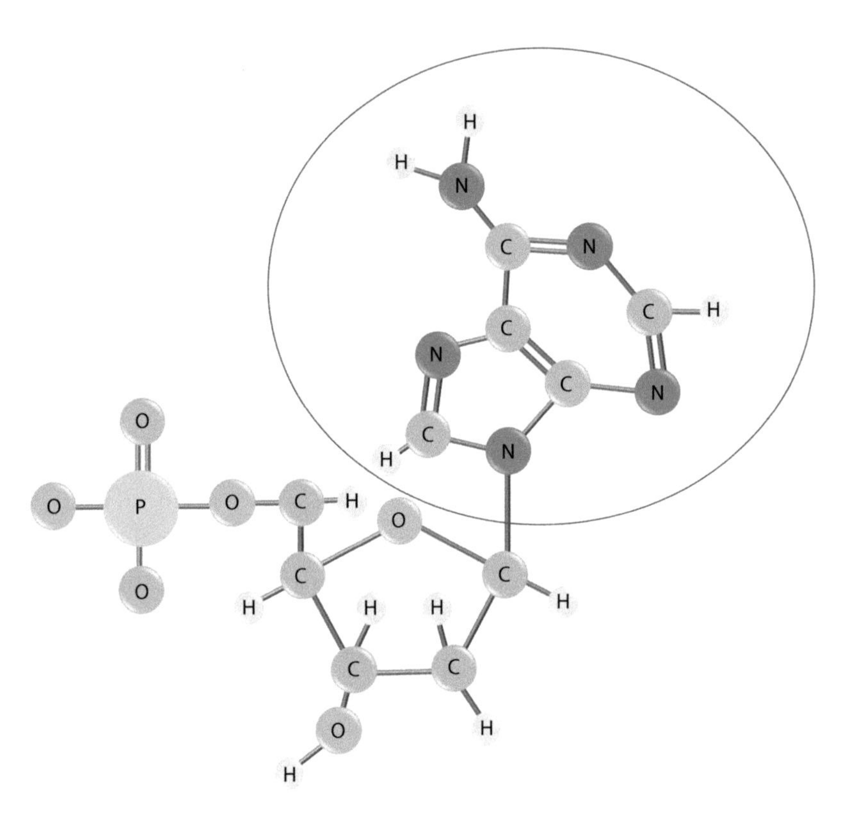

FIGURE 5.1
Structure of DNA. The circled region indicates the base portion of the nucleotide.

consists of nucleotides, which are structures with nitrogen containing rings (bases), sugar molecules, and phosphate ions (Rogers, Adams and Pennathur 2013). DNA possesses a double helix structure created from two nucleotide chains (Rogers, Adams and Pennathur 2013). The strands are held together by the attraction of bases found in the nucleotides (Figure 5.1). The bases adenine, guanine, cytosine, and thymine pair via hydrogen bonding and occur in a specific combination (Figure 5.2) (Rogers, Adams and Pennathur 2013).

5.2.2 DNA Origami

In the early 1980s, Ned Seeman, an X-ray crystallographer at New York University, spent more than a decade determining how to use DNA for the construction of self-assembling nanostructures (Tabata 2010; Service 2011). In 2006, Paul Rothemund, a chemist at the California Institute of Technology in Pasadena, developed DNA origami (Service 2011). This process involves using a 7000-base-pair DNA strand with a known sequence (Service 2011). Computer simulations allowed researchers to determine how to fold a single DNA strand to create an arbitrary shape (Service 2011). The process involves

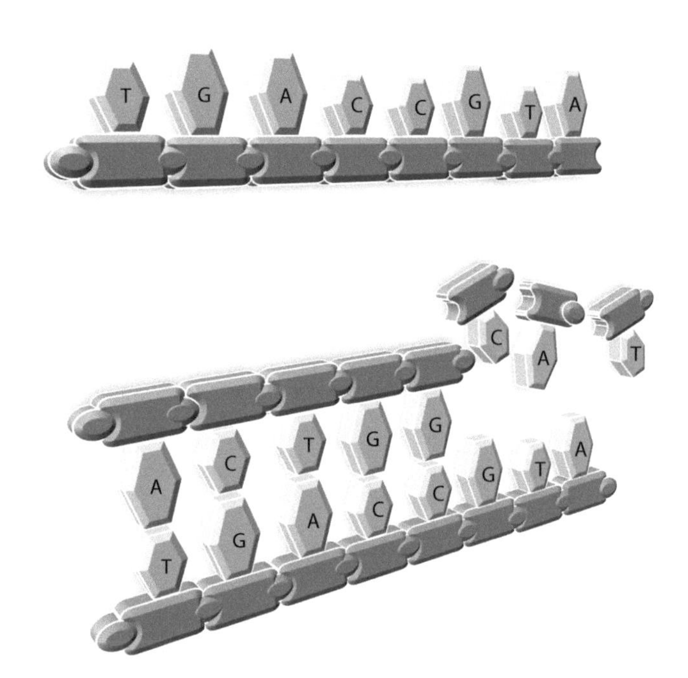

FIGURE 5.2
DNA strand formed from the hydrogen bonding between base pairs.

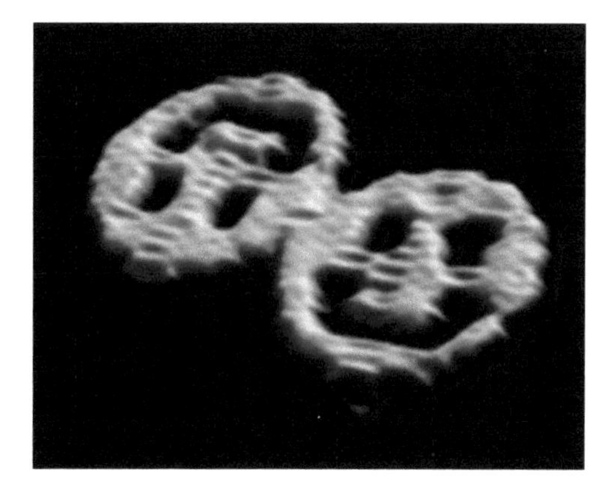

FIGURE 5.3
Atomic force microscope of DNA origami structure. Image originally created by Paul W. K. Rothemund and Nick Papadakis.

using short "staple" DNA strands, which bind to sections of longer strands. This binding encourages the longer strand to fold (Service 2011). The first shape created was a DNA smiley face, shown in Figure 5.3 (Service 2011). The site-directed binding of the staple strands allows long nucleotide strands to fold into arbitrary shapes tens to hundreds of nanometers in size (Figure 5.4)

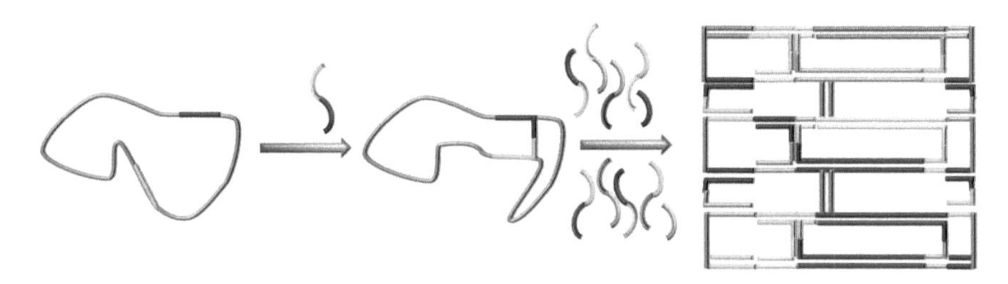

FIGURE 5.4
Folding of DNA into arbitrary shapes using shorter staple strands.

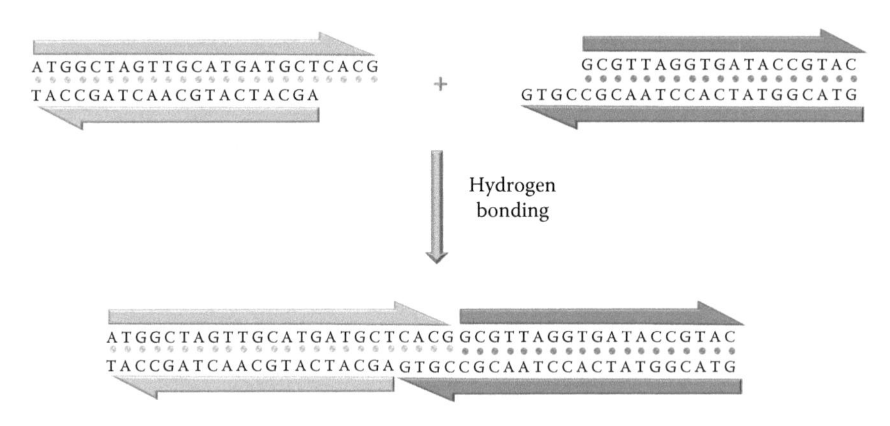

FIGURE 5.5
Sticky-ended cohesion between DNA strands.

(Tabata 2010). The molecular recognition properties of DNA strands promote the assembly of molecule-based structures with nanometer-scale precision (Michelotti et al. 2012). Bases found in DNA are effective, nanoscale building blocks because they follow simple, site-specific, and predictable attachment (Michelotti et al. 2012). DNA origami is driven by sticky-ended cohesion between staple strands and long DNA strands (Figure 5.5). This process is driven by hydrogen bonding and is used to combine pieces of DNA to make shapes (Neeman 2010).

5.3 Self-Assembled Monolayers

Self-assembled monolayers (SAMs) form when molecules spontaneously adsorb and arrange themselves on an appropriate substrate without intervention from an outside source (Colorado and Lee 2001). SAMs are organic thin-films that are 1–3 nm thick (Ulman 1996). SAMs are formed via self-assembly—the formation of nanostructures without external intervention

(Pradeep 2007). Irving Langmuir was the first to study monolayers in 1917 (Pradeep 2007). His study involved the observation of amphiphilic molecules, that is, molecules with polar and nonpolar characteristics on the surface of water (Pradeep 2007). Further study of molecular films was conducted by Katherine Blodgett who developed a technique for the transfer of monolayers onto solid supports (Pradeep 2007). In 1946, the spontaneous formation of molecules on a platinum surface was first observed (Pradeep 2007). The study of SAM grew significantly in 1983. At this time, researchers discovered that ordered, molecular films could be prepared on gold surfaces via solution adsorption (Pradeep 2007).

5.3.1 Alkanethiol Structure

Alkanethiols are frequently used as the molecular building blocks for self-assembled monolayers. As shown in Figure 5.6, alkanethiols contain three parts: a sulfur atom for attachment to a gold or silver metal surface, a carbon chain containing -CH_2- groups, and a functional head group. The sulfur atom and the carbon chains play an active role in the assembly of the SAM. The head group is used to change the surface chemistry of SAM. By simply changing the head group, a surface can be created which is hydrophobic, hydrophilic, and protein resistant, among other properties. This enables a researcher to custom design surfaces to serve any desired function (Jasty 2006).

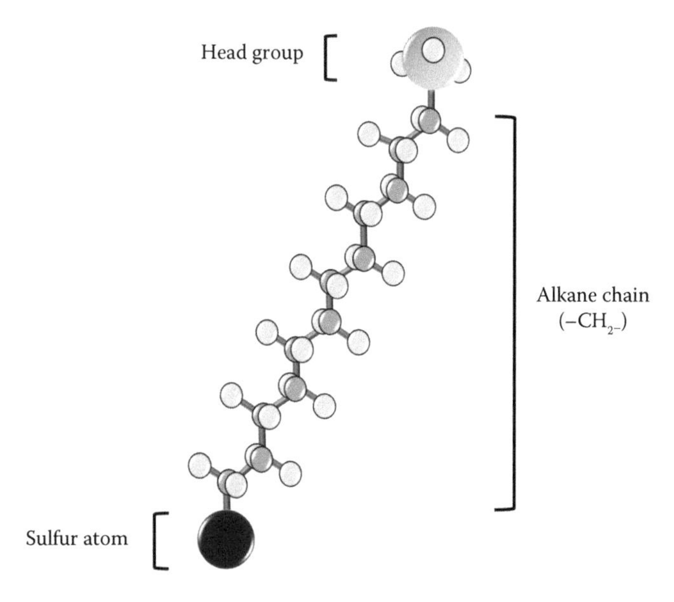

FIGURE 5.6
Diagram of an alkanethiol molecule.

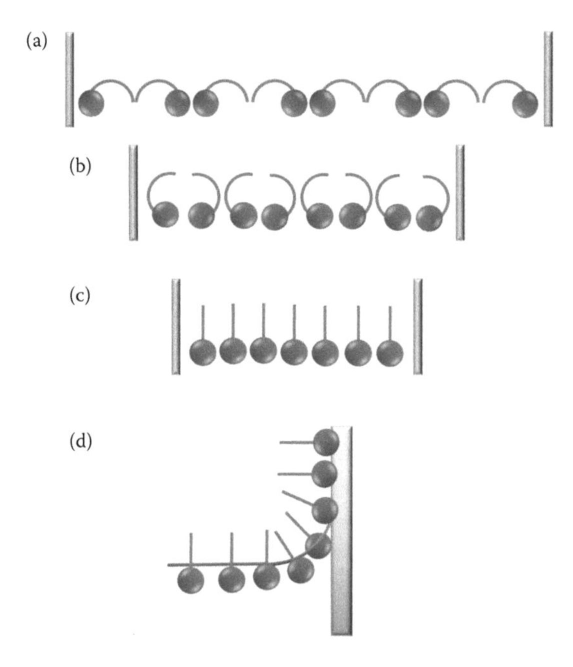

FIGURE 5.7
Langmuir–Blodgett technique. Disordered molecules (a) are compressed (b) to form ordered structures (c). Immersion of a substrate facilitates transfer of the film to a solid substrate (d).

5.3.2 Langmuir-Blodgett SAM Formation

Self-assembled monolayer formation using the Langmuir-Blodgett technique involves the spreading of amphiphilic molecules on the surface of a liquid (Pradeep 2007). Initially, the molecules are disordered (Figure 5.7a). Organization is accomplished via the application of pressure (Figure 5.7b); compression of the molecules across the surface of the liquid organizes the molecules (Figure 5.7c). Afterwards, a substrate is immersed in the liquid containing the ordered film, and the film is transferred to the surface of a solid substrate (Figure 5.7d) (Pradeep 2007).

5.3.3 SAM Assembly

SAMs on gold are the most thoroughly studied monolayer system because gold is inert, and alkanethiol molecules produce well-ordered SAMs on gold (Colorado and Lee 2001). Additionally, gold does not form an oxide at room temperature (Pradeep 2007). SAMs are formed by immersing substrates, such as gold, in dilute solutions of alkanethiols (Figure 5.8) (Pradeep 2007). Alkanethiol molecules form SAMs on gold substrates in stages, starting with the sulfur atom attaching to the gold surface via the formation of a stable, semicovalent bond (Canaria et al. 2006; Jasty 2006). This step occurs at a rapid rate and the attached molecules are disordered (Canaria et al. 2006; Pradeep 2007).

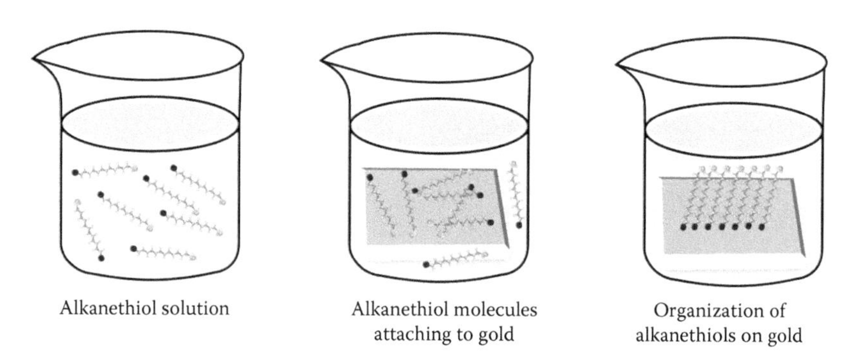

FIGURE 5.8
Formation of SAM via solution adsorption.

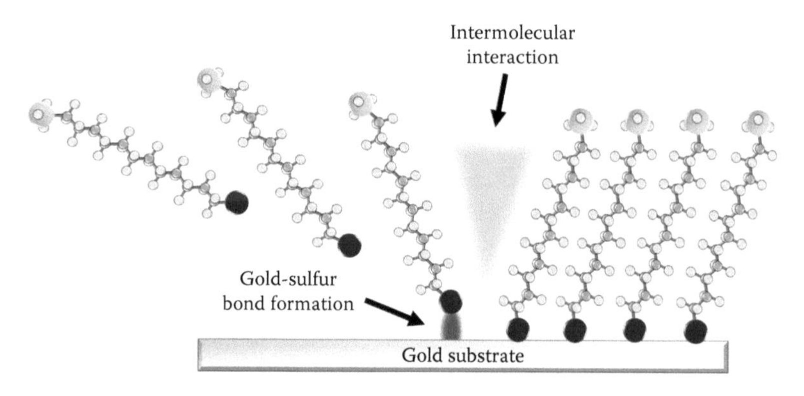

FIGURE 5.9
Chemical bonding and intermolecular interactions involved in SAM formation.

The next step involves hydrophobic London dispersion forces acting between neighboring carbon chains, as shown in Figure 5.9 (Jasty 2006). Even though London dispersion forces between the carbon chains are weaker than the interaction between the sulfur atom and the gold surface, these forces are primarily responsible for formation of a well-ordered SAM (Colorado and Lee 2001). The organizational step is slow, occurring over the course of 12–24 hours (Pradeep 2007). The rate of each step depends on experimental factors such as the concentration of the solution and the length of the carbon chain (Pradeep 2007). It is important to point out that the length of the alkanethiol carbon chain also affects how well the SAM is organized. With longer chain alkanethiol molecules, the London Dispersion forces acting between the carbon chains allows all chains to stand up, and reduces the number of defective sites in the final molecular film (Pradeep 2007). Defects appear in the resultant SAM when alkanethiol molecules with smaller carbon chains are used due to weaker London dispersion forces acting between neighboring molecules (Pradeep 2007).

Alkanethiol molecules attach to gold surfaces forming a close-packed hexagonal arrangement, as illustrated in green in Figure 5.10a. The unit cell

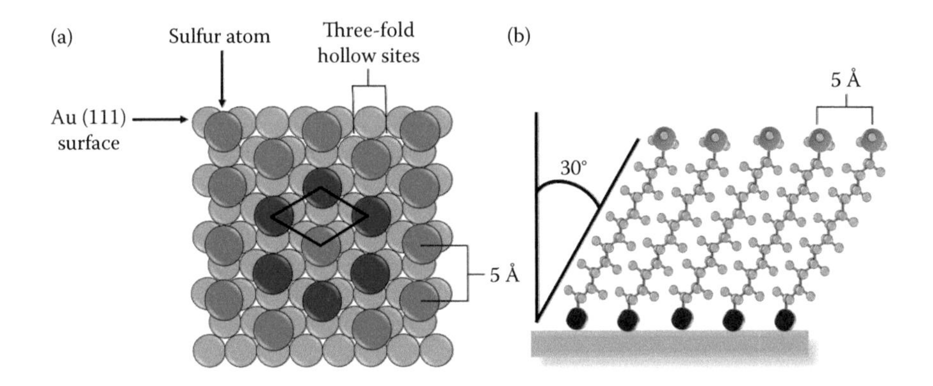

FIGURE 5.10
Arrangement of sulfur atoms of alkanethiol molecules on the surface of gold (a). Chain tilt of alkanethiol molecules assembled on gold (b).

formed when the sulfur atoms of alkanethiol molecules attach to gold exhibits a $\sqrt{3} \times \sqrt{3}R30°$ organization above the underlying gold substrate, as shown in Figure 5.10a. The sulfur atoms of the alkanethiol molecules sit in the threefold hollow sites of the underlying gold substrate. This organization of alkanethiols on gold results in a spacing of 5 Å between alkanethiol molecules that are tilted 30° with respect to a surface normal (Ulman 1996; Colorado and Lee 2001).

5.3.4 SAM Applications

SAM add functionality to nanometer-scale objects such as nanoparticles and nanowires. SAM functionality allows researchers to engineer surface properties and localize chemical reactions (Ulman 1996). For instance, SAMs can make a surface receptive to specific types of molecules or render surfaces chemically inert (Rogers, Adams and Pennathur 2013). Interestingly, octadecanethiol monolayers provide superb protection of the metal surface against oxidation. It is reported that silver surfaces coated with SAMs can be kept in ambient conditions without tarnishing for many months. Also, reports state coating copper surfaces with SAMs protects the metal when exposed to nitric acid (Ulman 1996). It is noteworthy to mention that SAMs may provide pathways to nanoscale devices for use in "organic electronics" (Whitesides 2005).

5.4 Dendrimers

The name dendrimer comes from the Greek word "Dendron," meaning "tree" (Baig et al. 2015). The name is appropriate because dendrimers are molecules with "tree-like" structures (Patel and Patel 2013). These molecules were first reported in 1974 by Fritz Vögtle (Kubiak 2014). Dendrimers are suitable for a variety of biomedical applications because they have sizes and physicochemical properties comparable to biomolecules, such as proteins (Agrawal and Kulkarni 2015).

5.4.1 Dendrimer Structure

The structure of dendrimers has a significant impact on their chemical and physical properties (Kubiak 2014). Dendrimers are usually 5 nm across (Rogers, Adams and Pennathur 2013). Due to the extensive network of branches in dendrimer molecules, many applications are possible including, catalysis, electronics, and drug release (Baig et al. 2015). Dendrimers contain three different regions: the core, branches, and terminal groups, as shown in Figure 5.11 (Nikalje 2015). The core is a single atom or molecule dendrons are attached to. The dendrons (branches) are monomers linked to the core, forming layers and building successive generations (Wilczewska et al. 2012). Branched monomers, dendrons, extend from the core. At the end of the branches are terminal groups, which can be easily modified to change the molecule's chemical and physical properties such as chemical reactivity, stability, solubility, and toxicity (Kubiak 2014). By selecting proper terminal groups dendrimers can be tailored for specific uses (Rogers, Adams and Pennathur 2013). For instance, the solubility of dendrimers is strongly influenced by the nature of the terminal groups. Hydrophilic functional groups increase the solubility of dendrimers in polar solvents, while dendrimers with hydrophobic functional groups are soluble in nonpolar solvents (Klajnert and Bryszewska 2001). Additionally, the internal cavities can be used as a "pocket" for small particles (Kubiak 2014).

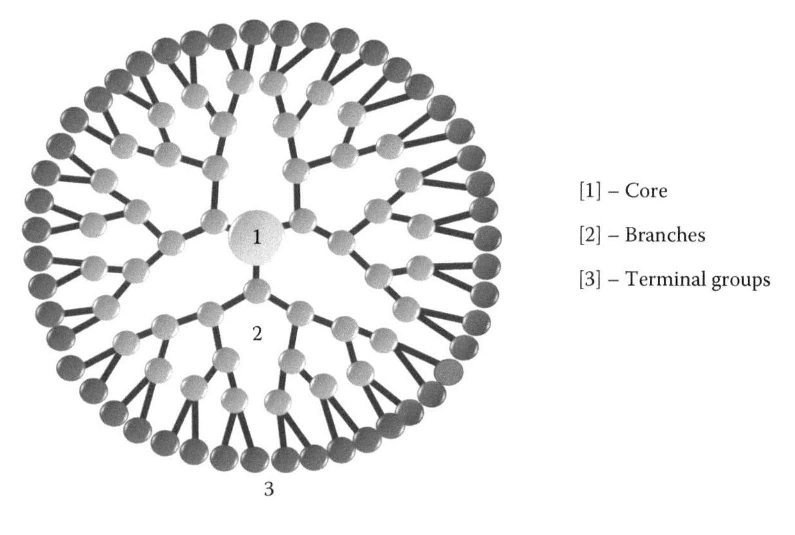

FIGURE 5.11
Structure of a dendrimer.

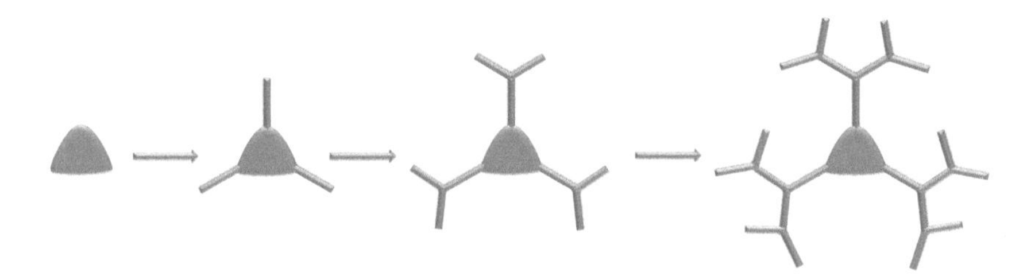

FIGURE 5.12
Divergent dendrimer synthesis.

5.4.2 Divergent Dendrimer Synthesis

The first reported dendrimer synthesis was the divergent method, shown in Figure 5.12. This method involves outward dendrimer growth by the successive attachment of molecules to the core (Baig et al. 2015). As the synthesis proceeds, successive layers are added, allowing dendrimers to expand to the desired size (Agrawal and Kulkarni 2015). Every new layer creates molecules with twice as many active end groups and produces a molecule that is twice the molecular weight of the dendrimer with the previous layer (Baig et al. 2015).

5.4.3 Convergent Dendrimer Synthesis

An alternative dendrimer synthesis is the convergent method (Figure 5.13). In this method, highly branched dendrons are first obtained, which are then connected to a core. This method is favorable because it offers better control over dendrimer growth. Additionally, this synthesis produces highly pure dendrimers, free of defects (Kubiak 2014).

5.4.4 Dendrimers in Medicine

Biocompatibility of dendrimers is determined by surface functional groups (Wilczewska et al. 2012). Dendrimers with appropriate functional groups can be added to drugs to enhance solubility (Agrawal and Kulkarni 2015).

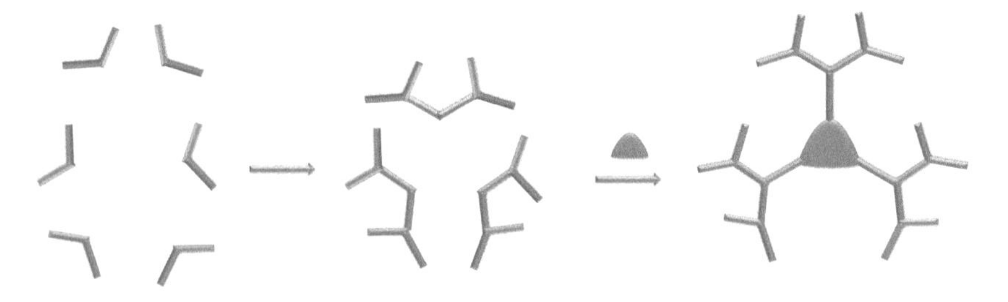

FIGURE 5.13
Convergent synthesis of dendrimers.

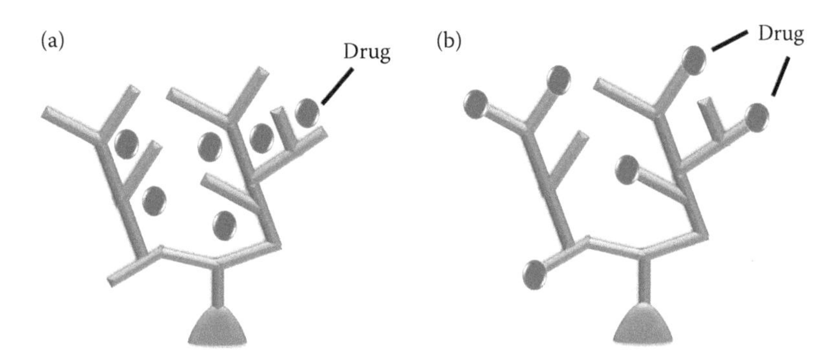

FIGURE 5.14
Dendrimer molecules with drug molecules encapsulated in branches (a). A dendrimer molecule with drug molecules at terminal branches (a).

Dendrimers can be modified using the empty spaces between branches, which are used to trap smaller molecules (Figure 5.14a) (Kubiak 2014). Synthetic macromolecules have been used to modify properties of molecules of particular interest in clinical, pharmaceutical, and biotechnical industries (Singh et al. 2014). For instance, functional groups can also be used to attach drug molecules to the dendrimer (Figure 5.14b). Functional group–drug interactions can occur via electrostatic or covalent bonds (Agrawal and Kulkarni 2015). Poly(amido amide) (PAMAM) is a dendrimer frequently used in biomedical applications. Both the structure of PAMAM dendrimers and the distribution of drugs or genes inside these molecules have been intensively investigated (Wilczewska et al. 2012).

5.5 Lipids

Lipids have a range of desirable properties for use in nanotechnology. Lipids can self-assemble into nanofilms and other nanostructures such as micelles, reverse micelles, and liposomes. Additionally, lipid assemblies can be attached to other nanostructures via specific chemical linkages. These features, along with transparency of lipid structures in visible light and their heat conductivity, have made lipids an important building block for nanotechnology (Mashaghi et al. 2013). Lipids involve molecules with a hydrophobic, hydrocarbon tail and a hydrophilic head group (Figure 5.15) (Wilson et al. 2002). Lipid molecules are 2–4 nm in size (Wilson et al. 2002). Addition of lipids to water results in the dissolution of the polar head groups in water. The hydrophobic tails do not dissolve; however, the tails stick together using weak London dispersion forces (Wilson et al. 2002). The shape of the resulting structures can be modified by changing the size of the head group and the length of the hydrocarbon tail (Wilson et al. 2002).

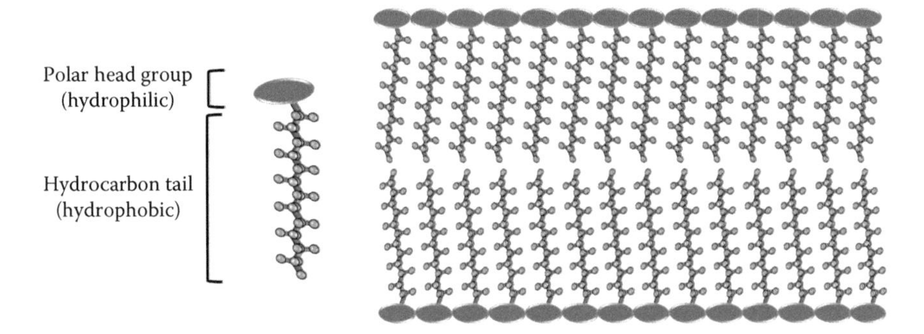

FIGURE 5.15
Lipid structure.

5.6 Micelles

Micelles are structures formed from the self-assembly of molecules known as surfactants (Rogers, Adams and Pennathur 2013). Surfactants are molecules with hydrophobic tails and hydrophilic head groups (Rogers, Adams and Pennathur 2013). The hydrophobic component of the micelle is a hydrocarbon chain, and the hydrophilic component is a polar group (Rogers, Adams and Pennathur 2013). When micelles are immersed in water, the hydrophilic carbon chains bunch together to minimize contact with water (Figure 5.16). This results in the polar, hydrophilic heads facing water (Rogers, Adams and Pennathur 2013). It is important to point out that spherical micelles closely resemble the structure of dendrimers (Rogers, Adams and Pennathur 2013).

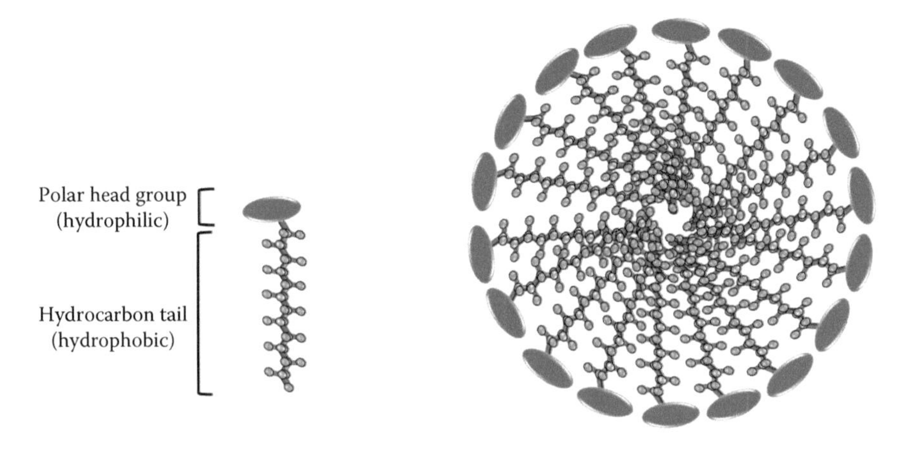

FIGURE 5.16
Structure of a spherical micelle.

A surfactant concentration in water that exceeds a value known as the critical micelle concentration (CMC) leads to the spontaneous generation of micelles 2–10 nm in diameter (Rogers, Adams and Pennathur 2013). Interestingly, bilayer micelles are the main component of cell membranes (Rogers, Adams and Pennathur 2013).

5.7 Molecular Electronics

Essentially all electronic processes in nature, from photosynthesis to signal transduction, occur in molecular structures (Heath and Ratner 2003). Molecular electronics can be defined as technology utilizing single molecules, small groups of molecules, or carbon nanotubes to perform electronic functions (Carroll and Gorman 2002). Such a scenario was suggested in a 1959 lecture by the eminent physicist and visionary, Richard Feynman:

> I don't know how to do this on a small scale in a practical way, but I do know that computing machines are very large; they fill rooms. Why can't we make them very small, make them of little wires, little elements–and by little, I mean little. For instance, the wires should be 10 or 100 atoms in diameter, and the circuits should be a few thousand angstroms across ... there is plenty of room to make them smaller (Carroll and Gorman 2002).

In 1974, Avi Aviram and Mark Ratner proposed the concept of using single molecules as diodes, electrical components allowing current to pass only in one direction (Carroll and Gorman 2002; Roth and Carroll 2004). Since then, molecular-scale electronics have attracted a growing interest due to possible applications in nanoelectronics (Heath and Ratner 2003). Molecular electronics involves the use of individual molecules acting as wires, switches, logic gates, transistors, and memory storage agents. Molecule-based electronics can contain one to a few thousand molecules per device (Rogers, Adams and Pennathur 2013). Each molecule is only a few billionths of a meter long (Kwok and Ellenbogen 2002). It is believed that the use of small organic molecules, carbon nanotubes, or biomolecules could lead to the creation of a computing system containing approximately 10 billion switches (Kwok and Ellenbogen 2002). Fabrication of molecular scale electronics involves arranging organic molecules on metal and semiconductor substrates (Vuillaume 2010). For electronics applications, molecular structures have four major advantages: (1) the size scale of molecules is between 1 and 100 nm; (2) intermolecular interactions can be exploited to form structures by self-assembly; (3) many molecules have multiple, distinct stable structures with distinct optical and electronic properties; and (4) it is possible to vary a molecule's transport, binding, optical, and structural properties (Heath and Ratner 2003). A key development came in the 1990s. During this time, researchers successfully measured the current flowing between two gold electrodes connected by

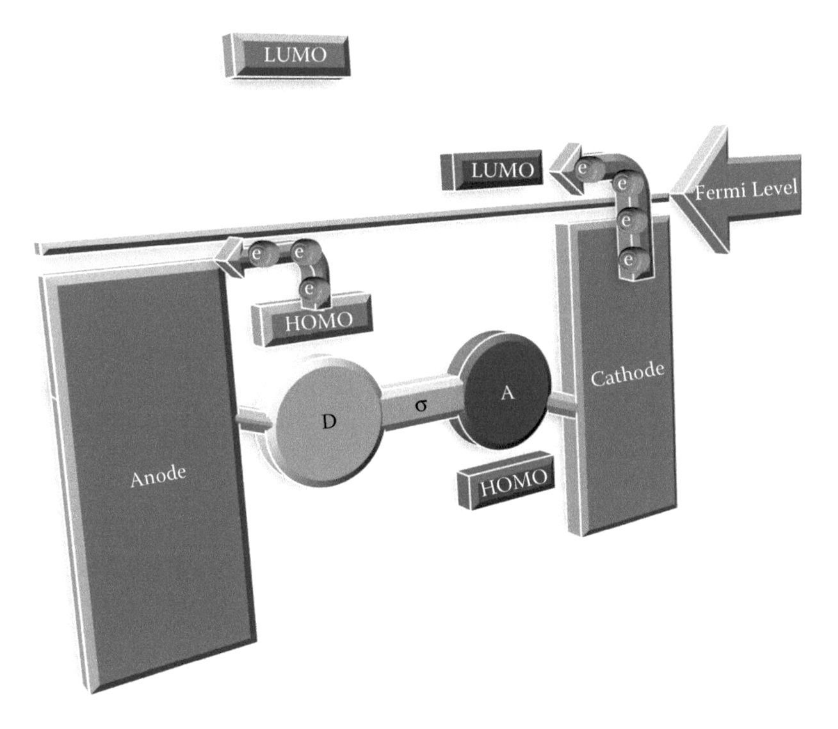

FIGURE 5.17
Electron transfer mechanism in molecule-based diodes.

a single molecule (Roth and Carroll 2004). Molecules used in molecular electronic applications have an electron donor-bridge-electron acceptor (DBA) structure (Heath and Ratner 2003). At positive bias voltages, the energy levels of the cathode align with the LUMO of the acceptor, allowing electric current to move from the metal cathode to the acceptor portion of the molecule, as shown in Figure 5.17. Charge flows between acceptor and donor when the energy levels of the metal anode are lowered enough to allow charge transfer from the HOMO of the donor to the anode. Since the HOMO/LUMO energies of the donor and acceptor portion of the molecule are close to the Fermi energies of the metals, conduction occurs under application of a low voltage (Carroll and Gorman 2002). The net result is that electrons flow from the cathode to the anode (Marruccio, Cingolani and Rinaldi 2004).

5.7.1 OLEDs

Molecule-based electronics are advantageous because of their low cost, light weight, mechanical flexibility, and excellent compatibility with plastic substrates (Wang et al. 2013). Advanced organic electronic systems are already utilized in commercial products with high efficiency, bright, and colorful thin displays (Forrest 2004). AMOLED (Active-Matrix Organic Light Emitting

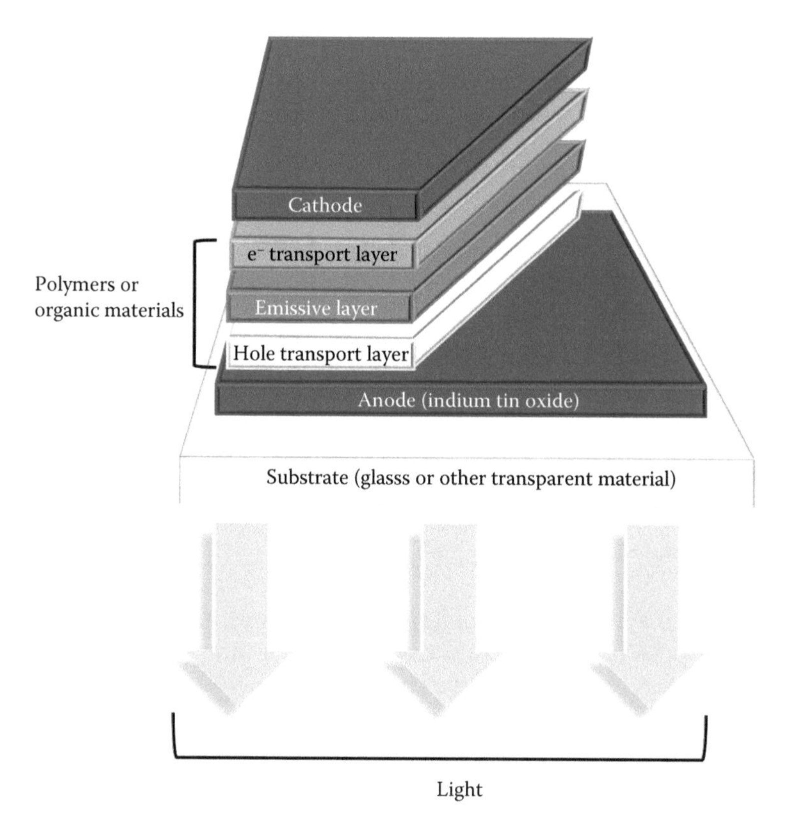

FIGURE 5.18
OLED structure.

Diode) display panels have replaced conventional LCD (Liquid Crystal Display) technology in smartphones (Chen et al. February 26–27, 2013). An organic light emitting diode (OLED) is an emissive device, 100–500 nanometers thick, consisting of an emissive layer sandwiched between two electrodes and deposited on a substrate (Karzazi 2014). OLEDs require a lower operating voltage and offer improved light output when compared to semiconductor-based LEDs (Karzazi 2014). Another attractive feature of organic electronics is the ability to deposit organic films on low-cost substrates such as glass, plastic or metal foils (Forrest 2004). This is due to the relative ease of processing of the organic compounds (Forrest 2004). These organic compounds can be tailored to optimize a particular function, such as charge mobility or luminescent properties (Forrest 2004). In three-layer OLEDs (Figure 5.18), the conductive layers consist of an electron transport layer (ETL) and hole-transport layer (HTL) (Karzazi 2014). In addition, there is an organic electroluminescent material consisting of small molecules or polymers that emit light when stimulated by electricity (Karzazi 2014). Conduction in the organic layer is driven by delocalization of π electrons over all or part of the organic molecule

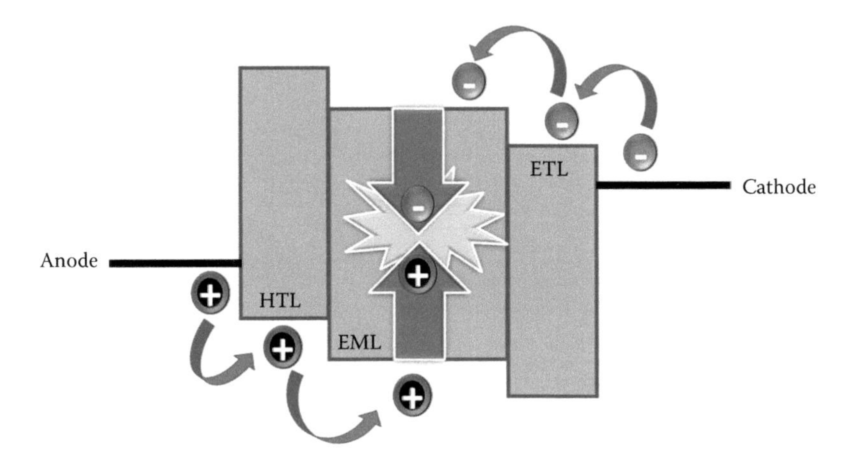

FIGURE 5.19
Injection of holes and electrons into the organic emissive layer where holes and electrons combine to produce light.

(Karzazi 2014). When voltage is applied, holes and electrons each migrate toward the oppositely charged electrodes via a hopping transport mechanism involving a series of "jumps" of the charge through the molecule (Figure 5.19). When an electron and hole are spatially close in the organic emissive layer (EML), they recombine. Upon recombination, energy is released as light. These processes result in very bright and crisp displays with power consumptions less than LCD and LED devices (Karzazi 2014).

5.7.2 Organic Solar Cells

Scientists have been intrigued with photoelectric molecules since the discovery of photography over 100 years ago (Hagfeldt and Gratzel 2000). Organic semiconductors such as dyes, dendrimers, oligomers, and polymers contain π electrons, which are mobile. Absorption of energy causes π bonds to break, resulting in the formation of excitons (Agrawal and Vivek 2014). The bandgap in organic semiconductors is tuned with the energy of the solar spectrum causing the absorption of photons producing electrostatically coupled electron–hole pairs called excitons. The excitons possess very small lifetimes of few picoseconds (10^{-12} s) and move within the molecules via a hopping mechanism (Agrawal and Vivek 2014). During the hopping mechanism, exciton dissociation occurs, which refers to the splitting of the electron–hole pair into free charges (Agrawal and Vivek 2014). Once the free charges are produced, they are collected at the electrodes. From there they are connected to the external circuit (Agrawal and Vivek 2014). Five processes need to be optimized to obtain high conversion efficiency of light into electrical energy. These processes include (1) absorption of light and generation of excitons, (2) diffusion of excitons to an active interface, (3) charge separation, (4) charge transport, and (5) charge collection. To create a working photovoltaic cell, photoactive materials are sandwiched between two metallic electrodes (of which one is transparent) to collect the photo-generated charges (Gupta 2015).

5.7.3 Dye-Sensitized Solar Cells (DSSCs)

Dye-sensitized solar cells (DSSCs) (also known as Gratzel cells) were first reported by Dr. Michael Graatzel in 1991. The components of DSSCs are shown in Figure 5.20. The cell consists of dye-coated titanium dioxide, an iodide/triiodide redox electrolyte, and a platinum or carbon counter anode (Zhang et al. 2009). The dye sensitizer absorbs the incident light and the light induces an electron transfer reaction (Wei 2010). At the heart of the device is the mesoporous oxide layer composed of a network of titanium dioxide (TiO_2) nanoparticles, which have been sintered together to establish electronic conduction (Hagfeldt and Gratzel 2000). Light harvesting occurs efficiently over the whole visible and near-IR range due to the very large internal surface area of the films (Hagfeldt and Gratzel 2000). Utilizing TiO_2 allows the effective surface area to be enhanced 1000-fold (Wei 2010). It was realized in the 1960s that molecular dyes, when excited, could inject electrons into the conduction band of the semiconductor substrates. In subsequent years, it was discovered that the dye could function most efficiently if chemisorbed on the surface of a semiconductor (Hagfeldt and Gratzel 2000). Titanium dioxide became the semiconductor of choice because it is cheap, abundant, nontoxic, and biocompatible (Hagfeldt and Gratzel 2000).

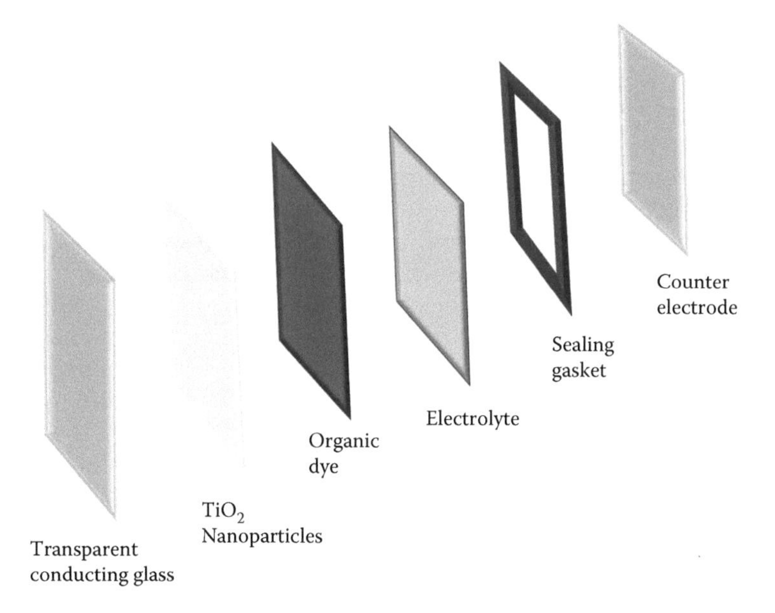

FIGURE 5.20
Components found in DSSCs.

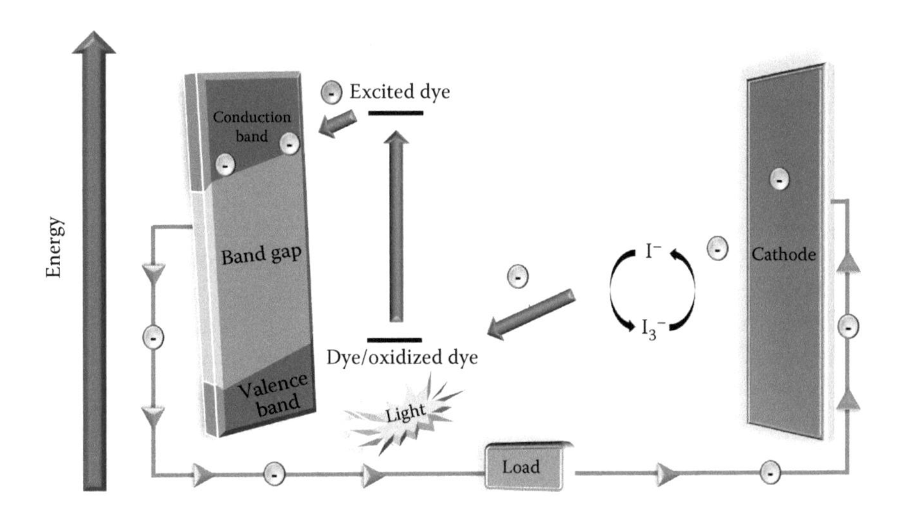

FIGURE 5.21
Electron transfer mechanism of DSSCs.

Wide bandgap semiconductors, such as TiO_2, may be sensitized to visible light using surface-adsorbed dye molecules (Cherepy et al. 1997). The energy conversion mechanism of DSSCs is illustrated in Figure 5.21. Excitation of the dye results in the injection of an electron into the conduction band of the titanium dioxide (Hagfeldt and Gratzel 2000). The original state of the dye is restored by electron donation from the electrolyte, which is usually an organic solvent containing the iodide/triioide (I^-/I_3) redox system. The iodide is regenerated via the reduction of triiodide at the counter electrode. An external load completes the circuit (Hagfeldt and Gratzel 2000). The light-driven process in DSSCs is regenerative (Cherepy et al. 1997). Overall, electric power is generated without permanent chemical transformation (Hagfeldt and Gratzel 2000). The operation of DSSCs parallels photosynthesis. The organic dye serves as the light harvester that produces electrons, titanium dioxide replaces carbon dioxide as the electron acceptor, iodide and triiodide replace water and oxygen as the electron donor and oxidation product, and a multilayer structure is used to enhance both the light absorption and electron collection efficiency (Cherepy et al. 1997). It is reported that the power conversion efficiency of DSSCs is approximately 11.5% (Wei 2010).

5.8 Drug Delivery Systems

Certain nanoscale materials have allowed biological tests to be performed quickly, with improved sensitivity and increased flexibility. With the help of nanotechnology early detection, prevention, improved diagnosis, and proper treatment is possible (Nikalje 2015). Recent developments in nanotechnology have shown nanomaterials to have a great potential as drug carriers (Wilczewska et al. 2012). For this reason, a variety of nanoparticles such as dendrimers and nanoporous materials find application in the medical field (Nikalje 2015). New complex drug delivery mechanisms are being developed allowing drugs to permeate cell membranes and move into the cell's cytoplasm, thereby increasing efficiency (Nikalje 2015). Due to their small sizes, molecule-based nanostructures exhibit unique physicochemical and biological properties (e.g., an ability to cross cell and tissue barriers), which make them favorable for biomedical applications (Wilczewska et al. 2012). Cell-specific targeting can be achieved by attaching drugs to individually designed molecular nanostructures (Wilczewska et al. 2012). In this technique, the required drug dose is used while side effects are lowered significantly because the drug is deposited into the morbid regions only (Nikalje 2015). Drugs having side effects due to triggering immune system responses can be wrapped in nanoparticle coatings, which prevent the immune system from recognizing and reacting to them (Jain et al. 2010). This highly selective approach can reduce costs and pain to patients because targeted medicine reduces the drug consumption and treatment expenses (Nikalje 2015). Nanotechnology-based drug delivery relies upon three factors: (1) efficient encapsulation of the drugs; (2) successful delivery of drugs to the targeted region of the body; and (3) successful release of that drug (Nikalje 2015). Additionally, nano-based systems allow delivery of insoluble drugs, which allows the use of previously rejected drugs, or drugs that were difficult to administer at one time (Jain et al. 2010).

5.9 End-of-Chapter Questions

1. All of the following are the main components found in the DNA double helix EXCEPT:

   ___ nucleotides

   ___ sugars

   ___ phosphate groups

   ___ all of the above are found in the DNA double helix

2. All of the following are used to form base pairs in the DNA double helix EXCEPT:

   ___ adenine

   ___ guanine

   ___ dextrose

   ___ cytosine

3. Base pairs used to form the double helix structure of DNA are held together by what?
   ___ ionic bonding
   ___ covalent bonding
   ___ hydrogen bonding
   ___ none of the above
4. What is the role of the "staple strands" in DNA origami?
   ___ to encourage unfolding of longer DNA strands
   ___ to encourage folding of longer DNA strands
   ___ to link together and form arbitrary nanostructures
   ___ to link together and form longer DNA strands
5. The driving force behind DNA origami, sticky-strand cohesion, is encouraged by:
   ___ dipole forces
   ___ London-dispersion forces
   ___ hydrogen bonding
   ___ covalent bonding
6. SAM are ____ nanometers thick.
   ___ 20–13
   ___ 9–11
   ___ 4–8
   ___ 1–3
7. The most frequently used building blocks to create SAM are:
   ___ dendrimers
   ___ alkanethiols
   ___ polymers
   ___ lipids
8. The _____ atom of an alkanethiol molecule attaches bonds to the metal surface during SAM formation.
   ___ carbon
   ___ hydrogen
   ___ sulfur
   ___ oxygen
9. _____ are responsible for organizing SAMs on metal surfaces.
   ___ sulfur atoms
   ___ carbon chains
   ___ head groups
   ___ hydrogen atoms

10. Which of the following forces is involved in the organization of SAMs?
    ___ hydrogen bonding
    ___ London dispersion forces
    ___ dipole forces
    ___ ionic bonding
11. The surface chemistry of the SAM is controlled using:
    ___ head group
    ___ carbon chain
    ___ sulfur atom
    ___ nature of the substrate
12. Formation of SAM by transferring a film on the surface of a liquid to a solid substrate is referred to as the _____ method.
    ___ solution adsorption
    ___ gas-phase adsorption
    ___ Langmuir–Blodgett
    ___ chemical vapor deposition
13. What can be used to change the chemical and physical properties of dendrimers?
    ___ core
    ___ branches (dendrons)
    ___ functional groups
    ___ diameter
14. Dendrimers are usually _____ nm in diameter.
    ___ 2
    ___ 5
    ___ 10
    ___ 15
15. What changes the chemical and physical properties of dendrimers?
    ___ functional groups
    ___ branches
    ___ core
    ___ none of the above
16. _____ dendrimer synthesis involves outward growth by successful attachment of molecules to the core.
    ___ convergent
    ___ divergent

17. ___ dendrimer synthesis involves the formation of highly branched dendrons before they are connected to a core.
    - ___ convergent
    - ___ divergent
18. True or False. Dendrimers can serve as drug carriers by encapsulating molecules in the branches and attachment of molecules at the functional groups.
    - ___ True
    - ___ False
19. Lipids are molecules that contain:
    - ___ a hydrophobic hydrocarbon tail only
    - ___ a hydrophilic headgroup only
    - ___ a hydrophobic hydrocarbon tail and a hydrophilic head group
20. When lipids are added to water, the _____ dissolve in water.
    - ___ hydrocarbon tails
    - ___ polar head groups
21. When lipids are added to water, the _____ stick together.
    - ___ hydrocarbon tails
    - ___ polar head groups
22. Spherical micelles closely resemble _____.
    - ___ SAMs
    - ___ dendrimers
    - ___ DNA strands
    - ___ staple strands
23. In the donor-bridge-acceptor (DBA) portion of a molecule, the acceptor molecule is connected to the _____ and the donor molecule is connected to the _____.
    - ___ cathode, anode
    - ___ anode, cathode
24. Molecules with DBA regions act as _______ when voltage is applied.
    - ___ capacitors ___ diodes
    - ___ resistors ___ transistors
25. All of the following are found in OLEDs except:
    - ___ HTL
    - ___ ETL
    - ___ emissive layer
    - ___ all of the above are found in OLEDs

26. All of the following are needed for successful organic photovoltaic operation except:
    ___ exposure of photovoltaic to 700 nm (red) light
    ___ generation of excitons
    ___ diffusion of excitons to an active interface
    ___ charge transport
    ___ charge collection
27. What is the purpose of the dye in a dye-sensitized solar cell (DSSC)?
    ___ to generate electrons when exposed to light
    ___ to serve as a conduit for electrons
    ___ to hold the titanium dioxide nanoparticles together
    ___ none of the above
28. What is the role of titanium dioxide (TiO_2) nanoparticles in dye-sensitized solar cells (DSSC)?
    ___ to serve as a conduit for electrons
    ___ to hold the indium tin oxide (ITO) plates together
    ___ to produce holes when exposed to light
    ___ to generate electrons when exposed to light
29. What is the role of the iodide ion (I^-) in dye sensitized solar cells?
    ___ to serve as a conduit for electrons
    ___ to provide excited dye molecules with electrons
    ___ to produce holes when exposed to light
    ___ to generate electrons when exposed to light

References

Agrawal, A., and S. Kulkarni. 2015. "Dendrimers: A new generation carrier." *Int. J. Res. Dev. Pharm. Life Sci.* 4: 1700–1712.

Agrawal, G. D., and K. A. Vivek. 2014. "Organic solar cells: Principles, mechanism, and recent developments." *Int. J. Res. Eng. Technol.* 3: 338–341.

Baig, T., J. Nayak, V. Dwivedi, A. Singh, A. Srivastava, and P. K. Tripathi. 2015. "A review about dendrimers: Synthesis, types, characterization, and application." *Int. J. Adv. Pharm. Biol. Chem.* 4: 44–59.

Canaria, C. A., J. So, J. R. Maloney, C. J. Yu, J. O. Smith, M. L. Roukes, S. E. Fraser, and R. Lansford. 2006. "Formation and removal of alkylthiolate self-assembled monlolayers on gold in aqueous solutions." *Lab. Chip.* 6: 289–295.

Carroll, R. L., and C. B. Gorman. 2002. "The genesis of molecular electronics." *Angew. Chem. Int. Ed.* 41: 4378–4400.

Chen, X., Y. Chen, Z. Ma, and F. C. A. Fernandes. February 26–27, 2013. "How is Energy Consumed in Smartphone Display Applications?" *ACM HotMobile'13*. Jekyll Island.
Cherepy, N. J., G. P. Smestad, M. Gratzel, and J. Z. Zhang. 1997. "Ultrafast electron injection: Implications for a photoelectrochemical cell utilizing an anthocyanin dye-sensitized TiO_2 nanocrystalline electrode." *J. Phys. Chem. B* 101: 9342–9351.
Colorado, R. Jr., and T. R. Lee. 2001. "Thiol based self-assembled monolayers: formation and organization." *Encyclopedia of Materials: Science and Technology*.
Forrest, S. R. 2004. "The path to ubiquitous and low-cost organic electronic appliances on plastic." *Nature* 428: 911–918.
Gupta, A. 2015. "Organic solar cells and its characteristics." *J. Material Sci. Eng.* 4: 1–2.
Hagfeldt, A., and M. Gratzel. 2000. "Molecular photovoltaics." *Acc. Chem. Res.* 33: 269–277.
Heath, J. R., and M. A, Ratner. 2003. "Molecular Electronics." *Phys. Today* 56: 43–49.
Jain, N., R. Jain, N. Thakur, B. P. Gupta, D. K. Jain, J. Banveer, and S. Jain. 2010. "Nanotechnology: A safe and effective drug delivery system." *Asian J. Pharm. Clin. Res.* 3: 159–165.
Jasty, S. 2006. "Molecular self-assembly." *Mater. Matters* 1: 1–19.
Karzazi, Y. 2014. "Organic light emitting diodes: Devices and applications." *J. Mater. Environ. Sci.* 5: 1–12.
Klajnert, B., and M. Bryszewska. 2001. "Dendrimers: Properties and applications." *Acta. Biochim. Pol.* 48: 199–208.
Kubiak, M. 2014. "Dendrimers—fascinating nanoparticles in the application of medicine." *Chemik* 68: 141–150.
Kwok, K. S., and J. C. Ellenbogen. 2002. "Moletronics—future electronics." *Mater. Today* 5: 28–37.
Marruccio, G., R. Cingolani, and R. Rinaldi. 2004. "Projecting the nanoworld: Concepts, results and perspectives of molecular electronics." *J. Mater. Chem.* 14: 542–554.
Mashaghi, S., T. Jadidi, G. Koenderink, and A. Mashaghi. 2013. "Lipid Nanotechnology." *Int. J. Mol. Sci.* 14: 4242–4282.
Michelotti, N., A. Johnson-Buck, A. J. Manzo, and N. G. Water. 2012. "Beyond DNA origami: The unfolding prospects of nucleic acid technology." *WIRES Nanomed. Bionanotechnol.* 4: 139–152.
Neeman, N. C. 2010. "Nanomaterials based on DNA." *Annu. Rev. Biochem.* 79: 65–87.
Nikalje, A. P. 2015. "Nanotechnology and its application in medicine." *Med. Chem.* 5: 81–89.
Patel, H. N., and P. M. Patel. 2013. "Dendrimer applications—A review." *Int. J. Bio. Sci.* 4: 454–463.
Pradeep, T. 2007. *Nano: The Essentials Understanding Nanoscience and Nanotechnology*. New Delhi: Tata McGraw-Hill.
Rogers, B., J. Adams, and S. Pennathur. 2013. *Nanotechnology—The Whole Story*. Boca Raton: CRC Press Taylor and Francis Group.
Roth, S., and D. Carroll. 2004. "Molecular-scale electronics." In *One-Dimensional Metals*, 193–210. Wiley-VCH Verlag GmbH and Co.
Service, R. F. 2011. "DNA nanotechnology grows up." *Science* 332: 1140–1143.
Singh, U., M. M. Dar, and A. A. Hashmi. 2014. "Dendrimers: Synthetic strategies, properties, and applications." *Orient. J. Chem.* 30: 911–922.

Tabata, O. 2010. "A closer look at DNA nanotechnology." *IEEE Nanotechnol. Mag.* 4: 13–17.

Ulman, A. 1996. "Formation and structure of self-assembled monolayers." *Chem. Rev.* 96: 1533–1554.

Vuillaume, D. 2010. "Molecular nanoelectronics." *Proc. IEEE* 98: 2111–2123.

Wang, P. C., L. H. Liu, D. A. Mengistie, K. H. Li, B. J. Wen, T. S. Liu, and C. W. Chu. 2013. "Transparent electrodes based on conducting polymers for display applications." *Displays* 34: 301–314.

Wei, D. 2010. "Dye sensitized solar cells." *Int. J. Mol. Sci.* 11: 1103–1113.

Whitesides, G. M. 2005. "Nanoscience, nanotechnology, and chemistry." *Small* 1: 172–179.

Wilczewska, A. Z., K. Niemirowicz, K. H. Markiewicz, and H. Car. 2012. "Nanoparticles as drug delivery systems." *Pharmacol. Rep.* 64: 1020–1037.

Wilson, M., K. Kannangara, G. Smith, M. Simmons, and B. Raguse. 2002. *Nanotechnology—Basic Science and Emerging Applications*. Boca Raton: Chapman and Hall CRC.

Zhang, W., D. Zhang, T. Fan, J. Gu, J. Ding, H. Wang, Q. Guo, and H. Ogawa. 2009. "Novel photoanode structure templated from butterfly wing scales." *Chem. Mater.* 21: 33–40.

6

Inorganic Nanomaterials

Key Objectives

- Gain familiarity with the synthesis, properties, and applications of:
  - Metal nanoparticles
  - Quantum Dots
  - Nanowires
- Understand the associated properties such as:
  - Quantum confinement
  - Surface plasmon resonance
  - Bohr-exciton radius
  - Fermi Energy

6.1 Introduction

Gold and silver nanoparticles have been used throughout history for aesthetic and medical purposes (Sreeprasad and Pradeep 2013). Mixtures of gold salts with molten glass were used by medieval artisans to produce tiny gold colloids exhibiting a ruby color; this process was used to add color to ceramics and pottery (Sreeprasad and Pradeep 2013). This chapter will examine the synthesis, properties, and application of metal nanoparticles, nanowires, and quantum dots.

6.2 Physical and Chemical Properties of Metal Nanoparticles

The physical and chemical properties of metallic nanoparticles are considerably different than for bulk metals. For instance, metal nanoparticles exhibit lower

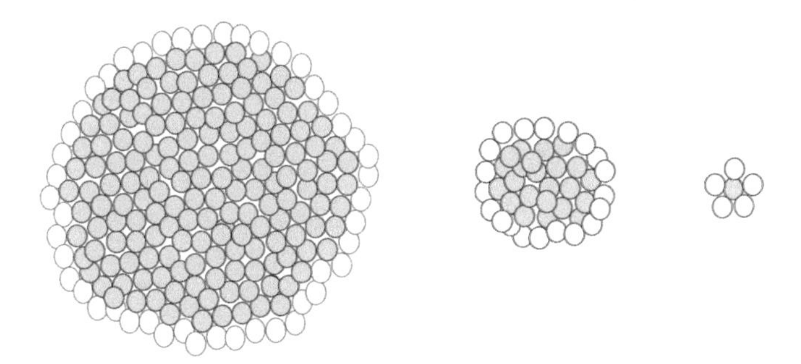

FIGURE 6.1
Increasing number of surface atoms with decreasing particle diameter.

melting points, higher surface areas, specific optical properties, and unusual mechanical strengths. Due to these unique properties, metal nanoparticles offer a variety of important uses in industrial applications (Horikoshi and Serpone 2013). Nanoparticle diameters can range between 1 and 100 nm. Nanoparticles possess a high surface to volume ratio, as illustrated in Figure 6.1. For comparison, bulk solid materials have less than 1% of their atoms on the surface, while nanoparticles have over 90% of their atoms on the surface. Due to the high surface to volume ratio, nanoparticles have proven to be more reactive catalysts in chemical reactions (Eustis and El-Sayed 2006; Rogers, Adams and Pennathur 2013).

6.2.1 Band Theory and Quantum Confinement

Possessing a few atoms in a material leads to other unique properties associated with nanoparticles (Rogers, Adams and Pennathur 2013). In bulk materials, atomic orbitals overlap producing bands (Figure 6.2a). In nanoscale metals orbitals are not continuous, they are spaced apart each with a representative energy (Figure 6.2b). This means the band gap of metal nanoparticles can be tuned by changing nanoparticle diameter (Rogers, Adams and Pennathur 2013).

As the particle size decreases, the electron becomes more confined in the particle (Eustis and El-Sayed 2006). As a result, valence bands (filled orbitals) and conduction bands (empty orbitals) break into quantized energy levels (Eustis and El-Sayed 2006). When the atomic orbitals of nanoparticles become discrete or quantized, quantum confinement occurs. Quantum confinement is the restriction of electron movement (Rogers, Adams and Pennathur 2013). Quantum confinement is achieved when electrons remain below the Fermi energy, E_f. Quantum confinement makes it possible to fabricate materials with tunable electrical properties, which can result in the development of novel electronic devices (Rogers, Adams and Pennathur 2013).

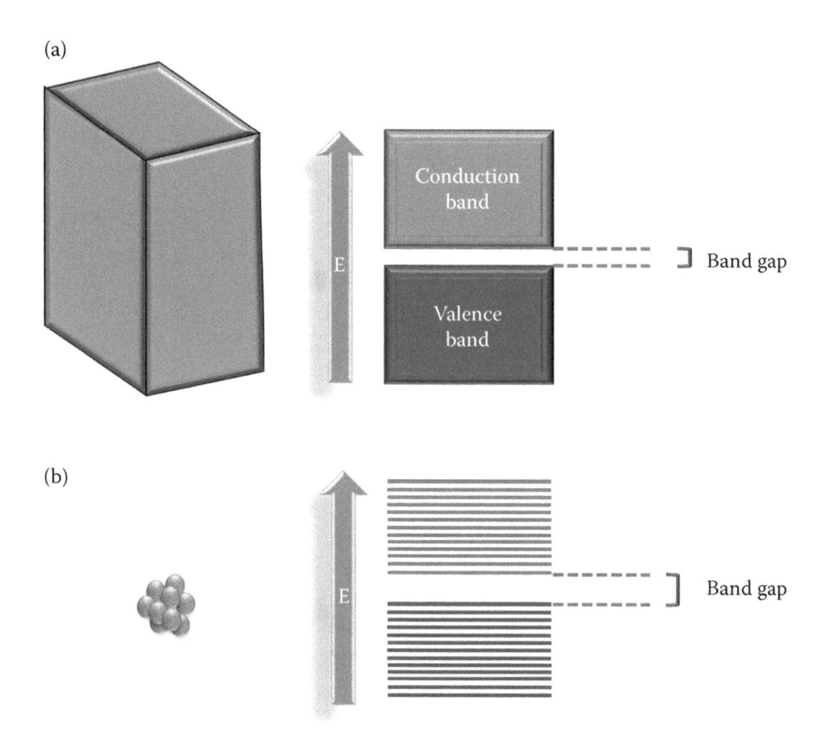

FIGURE 6.2
Continuous bands for bulk metals (a). The bandgap gets larger as the material gets smaller and orbitals in each band split apart (b).

6.2.2 Surface Plasmon Resonance and Optical Properties

The unique band structure of metal nanoparticles encourages very interesting light–matter interactions. When a metal particle is exposed to light, the electric field component of the electromagnetic wave induces collective, coherent oscillations of electrons in the nanoparticle (Figure 6.3) (Liz-Marzán 2004). This is referred to as surface plasmon resonance (SPR). During this phenomenon, the electric field displaces electron clouds (Noguez 2007). Additionally, the restoring force provided by the nuclei tries to compensate for the displacement, resulting in a unique resonance wavelength (Liz-Marzán 2004). At this wavelength, the amplitude of the oscillating electrons achieves a maximum. When a metal particle is exposed to light, the oscillating electromagnetic field of the light induces a collective coherent oscillation of the free electrons of the metal. This electron oscillation around the particle surface causes a charge separation with respect to the ionic lattice, forming a dipole oscillation along the direction of the electric field of the light (Huang and El-Sayed 2010). SPR induces a strong absorption of the incident light and thus can be measured using a UV–Vis absorption spectrophotometer. The SPR band is much stronger for noble metal nanoparticles, especially gold and

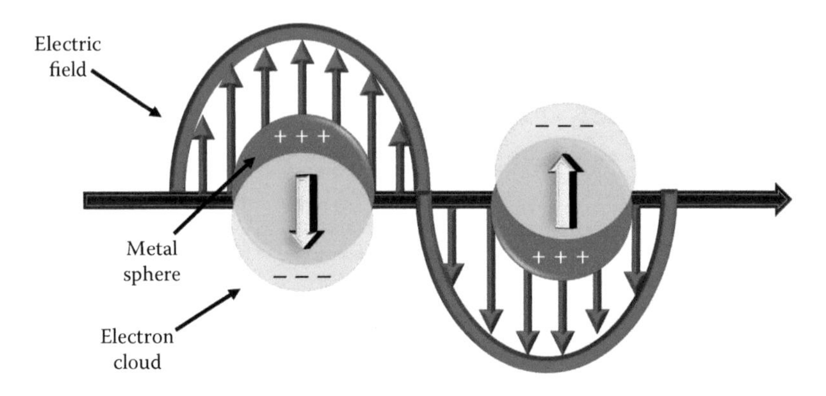

FIGURE 6.3
Illustration of localized SPR.

silver (Huang and El-Sayed 2010). The SPR band intensity and wavelength depends on the factors affecting the electron charge density on the particle surface such as the metal type, particle size, shape, structure, composition, and the dielectric constant of the surrounding medium. The interaction between the electromagnetic wave and metal nanoparticles results from two contributions: absorption and scattering. Light absorption results when the photon energy is dissipated due to inelastic processes. Light scattering occurs when the photon energy causes electron oscillations in the matter producing scattered light at the same frequency as the incident light (Rayleigh scattering), or at a shifted frequency (Raman scattering) (Huang and El-Sayed 2010). Due to the SPR oscillation, the light absorption and scattering are strongly enhanced (Huang and El-Sayed 2010). The oscillation wavelength associated with SPR usually occurs in the visible region for gold and silver nanoparticles (Eustis and El-Sayed 2006; Horikoshi and Serpone 2013; Kolwas, Derkachova and Jakubczyk 2016).

Gold and silver nanoparticles have a unique spectral response because specific wavelengths of light can allow electrons in the metal to collectively oscillate (NanoComposix - Plasmonics 2017). This oscillation results in the unusually strong scattering and absorption of visible light (NanoComposix - Color Engineering 2017; NanoComposix - Plasmonics 2017). When nanoparticles absorb light, the color perceived is opposite to the absorbed color (NanoComposix - Color Engineering 2017). For example, in colloidal form, 20-nm gold nanoparticles appear wine-red in color (Figure 6.4a) and silver nanoparticles appear yellowish-gray (Figure 6.4b) (Horikoshi and Serpone 2013). The optical properties of spherical gold nanoparticles are highly dependent on the nanoparticle diameter. As nanoparticle diameter increases, wavelengths absorbed shift towards longer wavelengths (known as red-shifting) (Huang and El-Sayed 2010). Manipulating nanoparticle size, shape, and composition makes it possible to tune the optical response from the ultraviolet, through the visible, and to the near-infrared regions of the

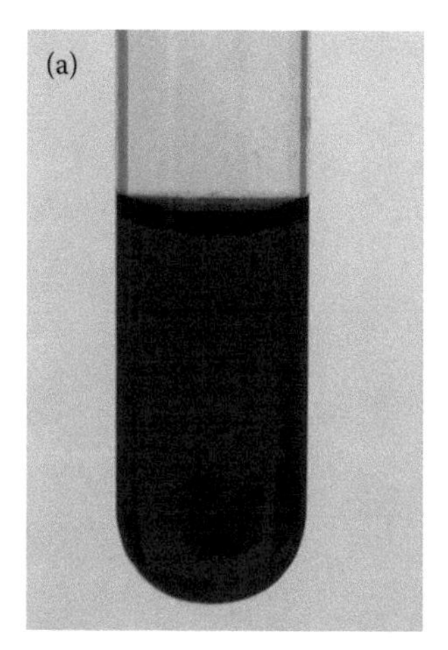

FIGURE 6.4
Gold (a) and silver (b) nanoparticles.

electromagnetic spectrum (NanoComposix - Plasmonics 2017). For example, the wavelength of the SPR band maximum of a spherical gold nanoparticle is 520–550 nm. When colloidal gold nanoparticles are irradiated with visible light, the visible light corresponding to the green color is absorbed and the particles now display a red purple color, which is the complementary color to green. In a colloidal silver nanoparticle solution, which has a plasmon resonance band maximum near 400 nm, the blue color of the visible light is absorbed and the silver particles now take on a yellow color—the complementary color to blue (Horikoshi and Serpone 2013).

6.2.3 Metal Nanoparticles and the Tyndall Effect

When illuminated, colloidal gold and silver nanoparticles scatter light (Kraemer and Dexter 1927). As shown in Figure 6.5a, a beam of the red light emanating from a laser pointer is visible in colloidal gold, and the green beam emanating from a green laser pointer is visible in colloidal silver (Figure 6.5b). This phenomena occurs because particles in colloidal solution are large enough to reflect or scatter light in all directions. Light scattering occurs when a light beam irradiates particles 1–200 nm in diameter present in a colloidal mixture (Yang 2013). In a true solution the dispersed particles are too small to scatter visible light (Yang 2013). This phenomena is referred to as the Tyndall effect.

6.2.4 Metal Nanoparticle Synthesis

One of the basic approaches used in the synthesis of colloidal gold involves the reduction of gold ions with an appropriate reducing agent (Turkevich 1985).

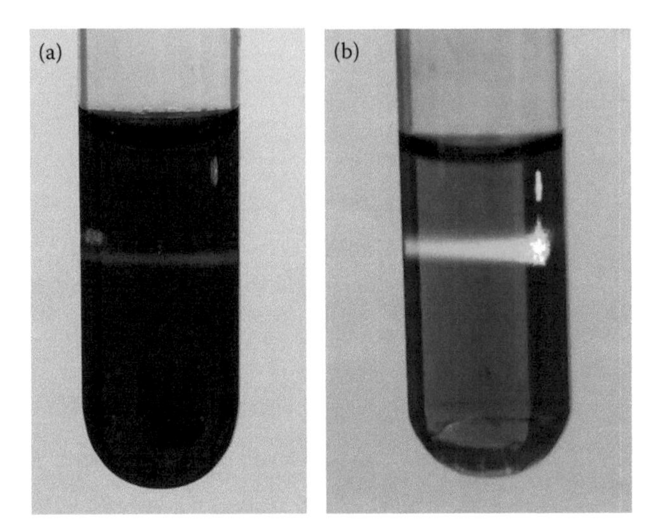

FIGURE 6.5
The Tyndall effect observed with gold (a) and silver (b) nanoparticles.

A common method of metal nanoparticle synthesis involves a solution-based reduction of a gold or silver ions using a citrate compound (Kimling et al. 2006). For instance, reducing chlorauric acid with sodium citrate at high temperatures (100°C) produces colloidal gold with a particle size of 20 nm diameter (Turkevich 1985). This method was first introduced by Turkevich, and has allowed the synthesis of spherical nanoparticles with tunable diameters (Kimling et al. 2006). An advantage of the Turkevich method is the seemless production of nanoparticles with shapes ranging from rods, wires, prisms, and plates (Horikoshi and Serpone 2013). This method allows manipulation of nanoparticle shape and size by varying experimental conditions such as the nature of the reducing agent, the time of reaction, and the temperature of the solutions (Horikoshi and Serpone 2013). The associated reaction for the citrate reduction of gold ions to form gold nanoparticles is shown in Equation 6.1 (Leng, Pati and Vikesland 2015):

$$2AuCl_4^- + C_6H_5O_7^{3-} + 2H_2O \rightarrow 2\,Au + 3CH_2O + 3CO_2 + 8Cl^- + 3H^+ \quad (6.1)$$

A similar process can be used to synthesize silver nanoparticles (Nain and Chauhan 2009). Silver nanoparticle preparation involves the reduction of silver ions using sodium borohydride or sodium citrate (Nain and Chauhan 2009). The chemical equation describing the reduction of silver ions to form silver nanoparticles is shown in Equation 6.2 (Nain and Chauhan 2009):

$$4Ag^+ + C_6H_5O_7Na_3 + 2H_2O \rightarrow 4Ag + C_6H_5O_7H_3 + 3Na^+ + H^+ + O_2 \quad (6.2)$$

It is noteworthy to mention that Michael Faraday used a similar approach to synthesize colloidal gold in the 1850s (Thompson 2007). Faraday employed a synthesis involving an aqueous solution of sodium chloroaurate ($NaAuCl_4$) and phosphorus in carbon disulfide. Faraday observed that the yellow color of sodium chloroaurate ($NaAuCl_4$) changed within minutes to a ruby color, the color characteristic of colloidal gold (Thompson 2007).

6.2.5 Metal Nanoparticles in Medicine

Colloidal gold nanoparticles have been used for a relatively long time for the treatment of diseases including cancer, rheumatoid arthritis, multiple sclerosis, and neurodegenerative conditions such as Alzheimer's disease (Jain et al. 2010). The advantages of using gold nanoparticles in medical applications include, ease of preparation, range of sizes, good biocompatibility, ease of functionalization, and ability to conjugate with other biomolecules without altering their biological properties (Jain et al. 2010). Also noteworthy, nanoparticles are designed to avoid the body's defense mechanisms (Nikalje 2015). Drugs may be also linked to the surface (Jain et al. 2010). Nanoparticles have a special property of high surface area to volume ratio, which facilitates attachment of various functional groups, allowing the nanoparticle to bind to tumor cells (Nikalje 2015). Once they are at the target site, the drug payload is released from the nanoparticle by diffusion, swelling, erosion, or degradation (Jain et al. 2010). Among various metals, silver and gold nanoparticles are of prime importance for biomedical use (Nikalje 2015). In ultrasound and MRI, nanoparticles are used as contrast agents (Nikalje 2015). Nanoparticles have been developed as effective target specific strategies for cancer treatment, as nanocarriers, and as active agents. Gold nanoparticles present tunable optical properties allowing the absorption of light at near UV to near infrared. Thus, nanoparticles heat up after irradiation of the body (or local area) with a magnetic field or another source of energy and consequently induce an increase in cancer cells' temperature until cell death (Silva, Fernandes and Baptista 2014). Reports state that nano shells of gold (120 nm diameter) were used to kill cancer tumors in mice by researchers at Rice University. The nano shells bond to cancerous cells by conjugating antibodies or peptides to the nano shell surface. The area of the tumor is irradiated with an infrared laser, which heats the gold sufficiently to kill the cancer cells (Nikalje 2015).

6.2.6 Kanzius Machine

John Kanzius was a broadcast engineer who, in 2003, developed a method to kill cancer cells with radio waves (Pitts 2008). Kanzius developed a special transmitter, which safely focused RF energy on tumors only. In 2005, Kanzius met Dr. Steven Curley, a professor at the M. D. Anderson Department of Surgical Oncology at the University of Texas. Dr. Curley found an interest in Kanzius' experiments. Dr. Curley believed if tumor cells absorbed

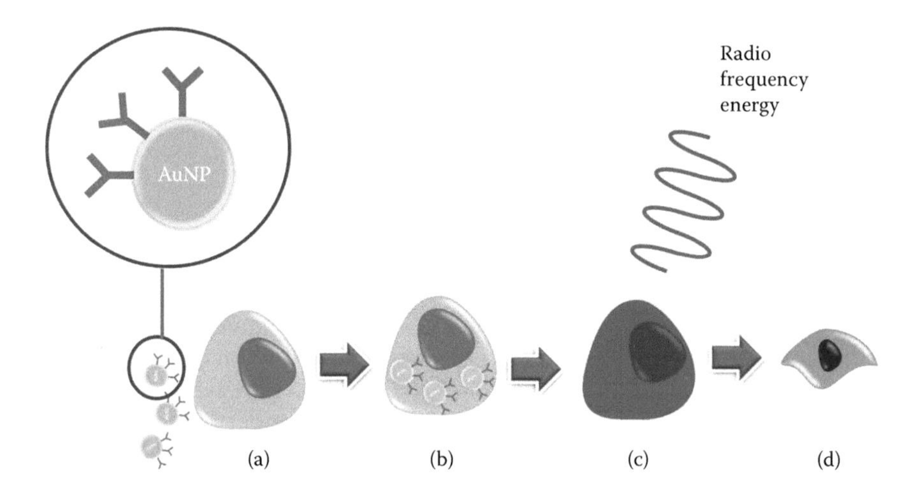

FIGURE 6.6
Noninvasive RF treatment. Gold nanoparticles are functionalized with antibodies (a), allowing the nanoparticles to penetrate tissues and reach the tumor (b). RF energy heats the nanoparticles (c) resulting in the death of the tumor (d).

nanoparticles, the particles would heat with RF energy, resulting in the destruction of the tumor. Kanzius created a special RF generator that targeted a tube of carbon nanotubes in a solution. Once the carbon nanotubes were exposed to RF energy, the solution began to boil (Raoof and Curley 2011). Soon after, animal testing was conducted at Rice University using rabbits. A solution containing carbon nanotubes was injected into cancerous tumors in rabbits with pancreatic or liver cancers. After injection they were exposed to a RF field for 2 minutes. The results were checked 48 hours later, and it was discovered that the tumors had been destroyed by heat with very little damage to neighboring tissues, as close as 2–5 mm away (Pitts 2008). This technique has been successfully used with gold nanoparticles (Figure 6.6), gold silica nanoshells, single-walled carbon nanotubes (SWNT), and water-soluble derivatives of C_{60} fullerenes (Raoof and Curley 2011).

6.3 Quantum Dots

The first evidence of quantum dots was observed in the early 1980s. This occurred when unusual optical behavior was obtained from glass samples containing cadmium sulfide or cadmium selenide subjected to high temperatures. Scientists concluded heating caused nanocrystallites of the semiconductor material to form, resulting in the unusual optical behavior (Reed 1993). Quantum dots are nanometer-scale semiconductor crystals with physical dimensions smaller than the Bohr-exciton radius (Jamieson et al. 2007). Quantum dot diameter is usually below 10 nm (Drbohlavova et al. 2009). Quantum dots contain three main parts (Figure 6.7). The central part of the quantum dot is known as the core. Cores are usually composed of elements from groups II and VI of the periodic table. Inner shell materials consist of semiconductor materials with a large band gap (Azzazy, Mansour and Kazmierczak 2007). The shell enhances the photostability of the quantum dot and improves its optical properties. Quantum dots have exceptional photophysical properties. Quantum dots are more photostable than organic dyes (Jamieson et al. 2007). Photostability, the ability of a fluorescent molecule to fluoresce under constant application of UV light, is an important feature in fluorescence applications. Organic fluorophores bleach after only a few minutes on exposure to light, however quantum dots can undergo repeated cycles of excitation and fluorescence for hours with a high level of brightness (Jamieson et al. 2007).

6.3.1 Quantum Dot Synthesis

There are two general approaches for the preparation of quantum dots: (1) formation of nanosized semiconductor particles through colloidal chemistry, and (2) epitaxial growth and/or nanoscale patterning (Drbohlavova et al. 2009). Usually, quantum dot synthesis relies on rapid injection of semiconductor precursors into hot and vigorously stirred organic solvents with molecules that can attach to the surface of the quantum dot. This synthesis is facile and can be performed in "one-pot," as reported in many sources (Drbohlavova et al. 2009). Lithographic preparation of quantum dots involves the combination of high-resolution electron beam lithography and subsequent etching. However, lithographic methods and subsequent processing leads to contamination, defects, and size nonuniformity. To add, patterning

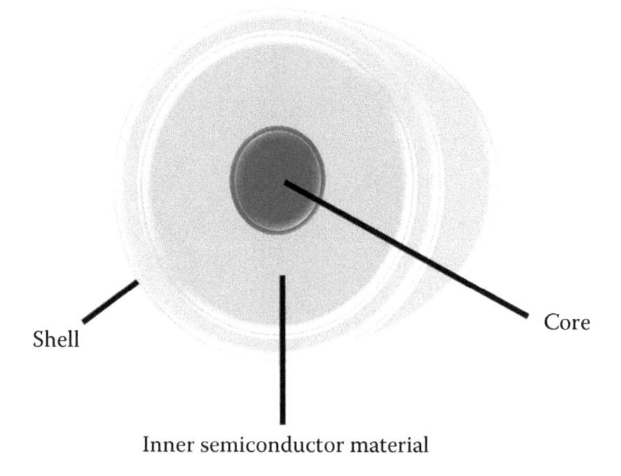

FIGURE 6.7
Structure of quantum dot.

methods of quantum dot synthesis are time-consuming and expensive processes (Drbohlavova et al. 2009). When quantum dots are synthesized, they should exhibit monodispersity (uniform sizes), possess the ability to be functionalized, highly crystalline, and defect-free (Pradeep 2007). Quantum dots can be synthesized in a reaction vessel (Pradeep 2007). Reactions can occur in mixtures of water and oil; oil serves as the primary phase and water as the secondary phase. Metal ions and organometallic materials can be added to the mixture resulting in the formation of the quantum dots in accordance to Equation 6.3 (Pradeep 2007):

$$CdCl_2 + Se(SiMe_3)_2 \rightarrow CdSe + 2Me_3SiCl \tag{6.3}$$

Another common quantum dot synthesis involves the use of molecular precursors (Pradeep 2007). This method involves the preparation of nanocrystalline seeds in a material controlling the growth of particles via chemical coordination (Pradeep 2007). Chemical coordination involves the binding of ligands (atoms, ions, or molecules) to metal centers where ligand atoms donate electrons to the metal centers. Trioctylphosphine oxide (TOPO) is used exclusively for this method of synthesis (Pradeep 2007). TOPO is used because it coordinates with inorganic materials and can undergo an exchange reaction with other ligands (Pradeep 2007). As the reaction proceeds, the particles grow to form monodisperse structures (Pradeep 2007). This method involves the injection of metal ion precursors into hot TOPO with continuous stirring (Figure 6.8) (Pradeep 2007). Oftentimes, a compound that contains the constituent elements and semiconductors is used in this type of synthesis (Pradeep 2007). For instance, $Cd(S_2CNEt_2)_2$ is used to produced cadmium sulfide (CdS) quantum dots (Pradeep 2007).

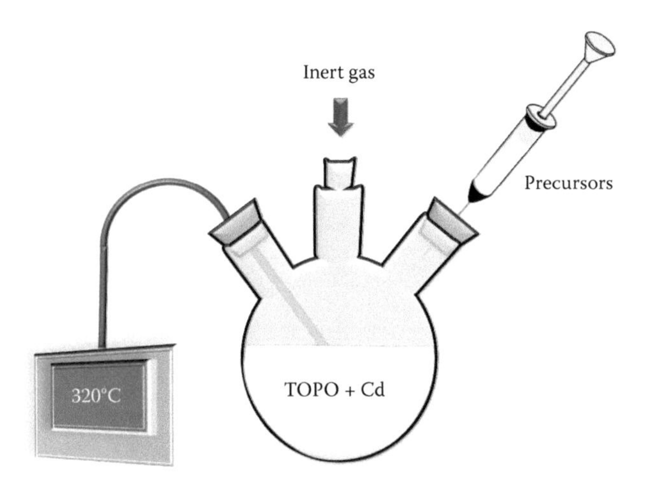

FIGURE 6.8
"One-pot" synthesis of quantum dots.

6.3.2 Optical Properties of Quantum Dots

The band structure of quantum dots is considerably different than the band structure of macroscopic materials (Pradeep 2007). In macroscopic systems, the valence band is occupied with electrons and the conduction band is vacant (Pradeep 2007). A separation exists between the valence band and the conduction band, referred to as a band gap, with a specific energy value (Eg) associated with it (Pradeep 2007). Excitation of an electron in these materials results in the formation of an electron–hole pair (Pradeep 2007). This pair is referred to as an exciton with a specific Bohr-exciton radius (Pradeep 2007). To exploit the unique optical properties of quantum dots, the diameter of the quantum dot is made comparable to the size of the Bohr-exciton radius (Pradeep 2007). The band gap of a quantum dot increases with a decrease in the Bohr-exciton radius (Pradeep 2007). As a result, colors ranging from blue to red can be emitted by the quantum dot (Pradeep 2007). Reports state that changing the diameter of cadmium selenide (Cd-Se) quantum dots from 11.5 to 1.2 nm produces a change of the quantum dot band gap of 1.8–3 eV. This corresponds to an emission of light from red to blue (Pradeep 2007). When quantum dots are exposed to ultraviolet light with a higher energy than that of the bandgap, an exciton (electron–hole pair) is formed because a valence band electron is promoted to the conduction band. When the electron returns to a lower energy level, light emission occurs (Azzazy, Mansour and Kazmierczak 2007). The electronic transition from the conduction band to the valence band is a function of the diameter of the quantum dot (Stokes, Stiff-Roberts and Dameron 2006). The energy gap increases with a decrease in the size of the nanocrystal, thus yielding a size-dependent rainbow of colors (Figure 6.9) (Klimov 2003). Quantum dots have been produced with

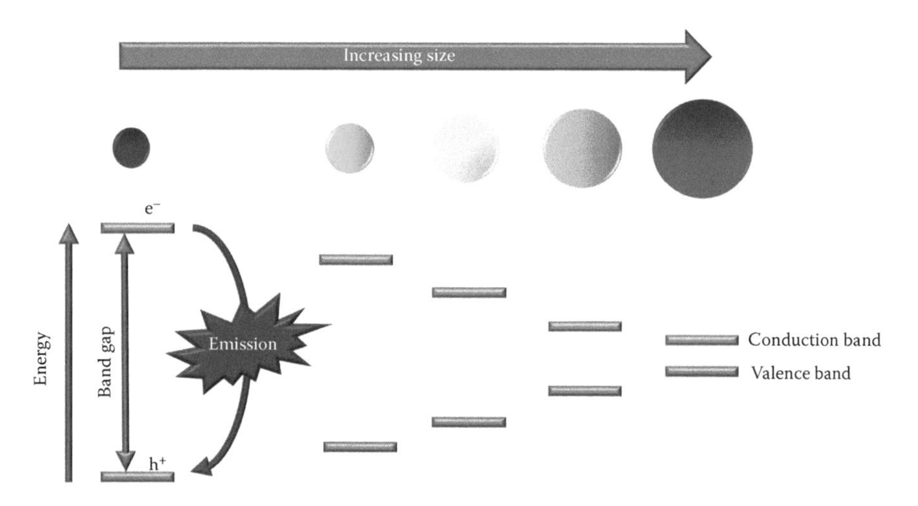

FIGURE 6.9
Size-dependent effects on bandgap and emission color.

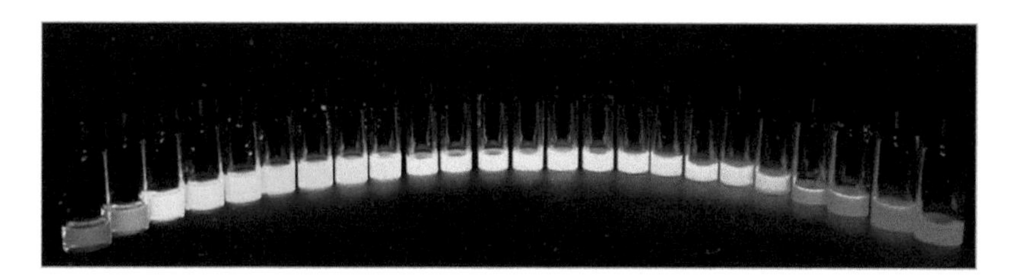

FIGURE 6.10
Quantum dot light emission. (Image originally created by Kenny Chou, courtesy of the Dennis Lab at Boston University.)

peak optical emission wavelengths ranging from the ultraviolet, through the visible (Figure 6.10), and into the infrared (Stokes, Stiff-Roberts and Dameron 2006). The emission of cadmium-selenium (CdSe) quantum dots, for example, can be tuned from deep red to blue by a reduction in the dot radius from 5–0.7 nm (Klimov 2003). As the particle gets smaller, the band gap gets larger. The band gap is so large that only high-energy ultraviolet light can excite electrons into the conduction band (Brown et al. 2012).

Semiconductor materials, such as quantum dots, emit light when exposed to photons with energies larger than the energy of the band gap of the material (Brown et al. 2012). This process is called *photoluminescence*. A valence-band electron absorbs a photon and is promoted to the conduction band. If the excited electron then falls back down into the hole it left in the valence band, it emits a photon having energy equal to the band gap energy. In the case of quantum dots, the band gap is tunable with the crystal size, and thus all the colors of the rainbow can be obtained from just one material (Brown et al. 2012).

6.3.3 Single Electron Transistor

It is possible for quantum dots to be used to create single electron devices (Pradeep 2007). The single electron transistor (SET) is a new type of switching device using controlled electron tunneling to switch current on and off (Gupta 2013). Tunneling is used to put a charge in a quantum dot surrounded by an insulating barrier (Pradeep 2007). Single-electron devices are promising, as new nanoscale devices, because they can control the motion of a single electron (Gupta 2013). The first single electron device was tested by Bell Laboratories in 1987 by Fulton and Dolan. Since then, the fabrication of SETs has become more and more sophisticated, and now allows for operation at room temperature. A SET consists of a small conducting island coupled to source and drain leads by tunnel junctions, which are capacitively coupled to a gate electrode (Figure 6.11) (Kumar and Kaur 2010). Due to the small size of the dot the electronic energy levels are discrete. This condition ensures that only one electron at the time is tunneling in or out of the quantum dot (Thiele 2016).

A gate voltage (Vg) is used to control the opening and closing of the SET, or in other words, it controls the one-by-one electron transfer (Gupta

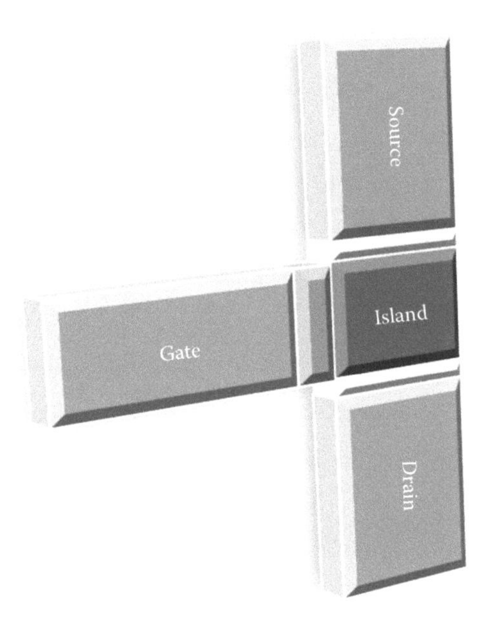

FIGURE 6.11
Single electron transistor (SET) structure.

2013). When there is no bias on any electrode, electrons in the system do not have enough energy to tunnel through the junctions (Kumar and Kaur 2010). In the OFF mode, it is not energetically favorable for source electrons to tunnel to the quantum dot (Figure 6.12a). The SET is in the ON mode when the proper gate voltage is applied (Figure 6.12b). This voltage lowers the quantum dot energy, encouraging an electron to tunnel to the quantum dot from the source electrode. The process continues with the potential energy of the quantum dot rising allowing the electron to tunnel from the dot to the drain (Figure 6.12c). Recent research in SET produces new ideas, which are going to revolutionize the random-access memory and digital data storage technologies. Single-electron transistor (SET) is a key element of current research area of nanotechnology, which can offer low power consumption and high operating speed (Gupta 2013).

6.3.4 Quantum Dots in Medicine

Cadmium selenide nanoparticles, in the form of quantum dots, are used in the detection of cancer tumors because they emit light when exposed to ultraviolet light. This facilitates easy removal of tumors since quantum dots possess size-tunable light emission. They can be used in conjunction with magnetic resonance imaging to produce exceptional images of tumor sites. As compared to organic dyes, quantum dots are much brighter and need only

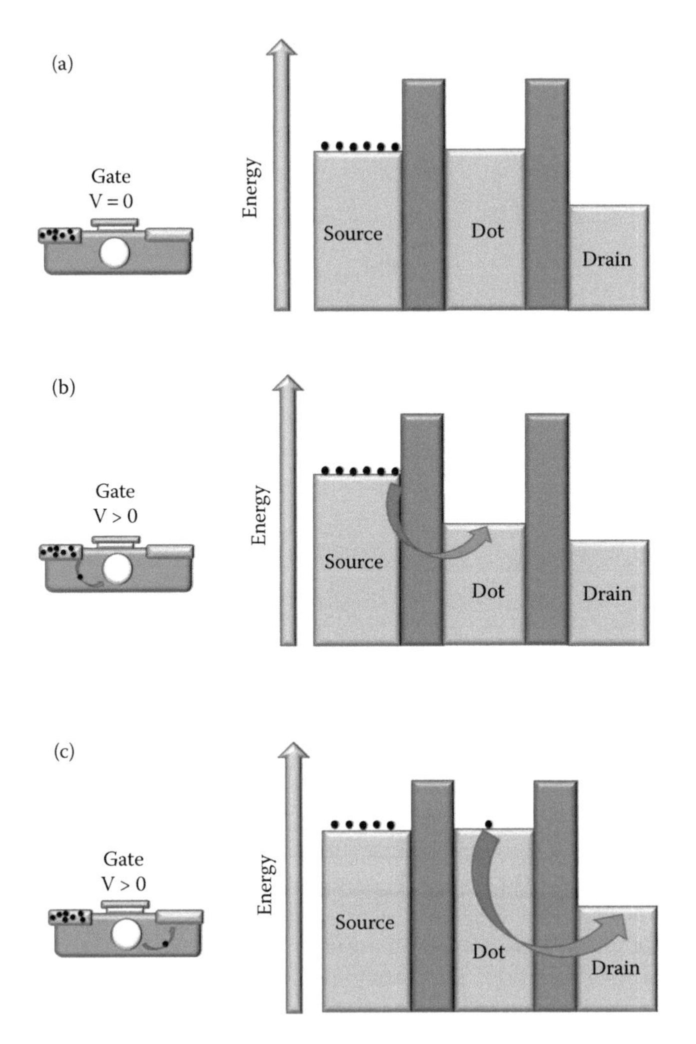

FIGURE 6.12
SET in the OFF state (a) and in the ON state (b). Tunneling from the quantum dot to the drain (c).

one light source for excitation. The use of quantum dots produces a higher contrast image at a lower cost than organic dyes routinely used as contrast media (Nikalje 2015).

6.4 Metal Nanowires

Metal nanowires have unique magnetic, optical, electrical, and catalytic properties. For this reason, they have potential applications in nanoelectronics

(Zhang et al. 2001). For instance, nickel nanowires are being studied because they have the potential to be used in hard drives in order to increase magnetic storage density (Bentley et al. 2005). Additionally, metal nanowires have potential use as active components or in interconnects in fabricating electronic, photonic, and sensing devices (Sun and Xia 2002). Additionally, they are of interest because of their high conductivity; this has resulted in the development of metal nanowire networks and meshes to use as an alternative to fragile and expensive indium tin oxide (ITO) (Groep et al. 2015). The high conductivity of silver nanowires also makes them promising candidates in flexible electronics (Liu and Yu 2011).

6.4.1 Polyol Synthesis of Silver Nanowires

The polyol synthesis is a large scale, solution-phase synthesis of silver nanowires. The polyol process generates silver nanowires by reducing silver nitrate with ethylene glycol (Sun and Xia 2002). Reduction is a chemical process involving the gain of electrons by an atom or ion. In the polyol process, silver ions (Ag^+) gain electrons to become neutral metal atoms. The polyol process is conducted in the presence of polyvinylpyrrolidone (PVP) (Sun and Xia 2002). PVP attracts silver ions (Ag^+) to polar groups (charged regions) on the PVP molecule (Figure 6.13). As silver ions congregate along the long carbon chain of PVP, nanoscale silver particles form in the presence of a reducing agent. As the process continues, the silver particles continue to grow and become silver nanowires. In this process, PVP acts as a directing agent assisting in the formation of a 1D silver nanowire structure (Zhu et al. 2011).

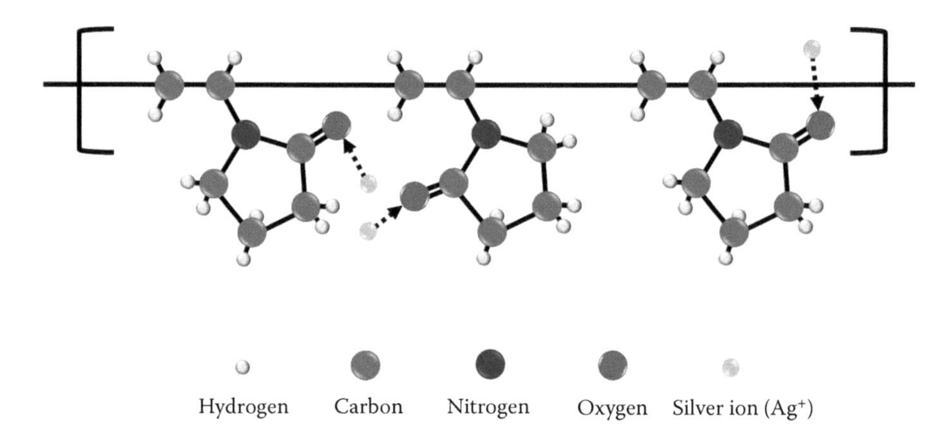

FIGURE 6.13
Association between silver ions and polar (charged) groups of PVP.

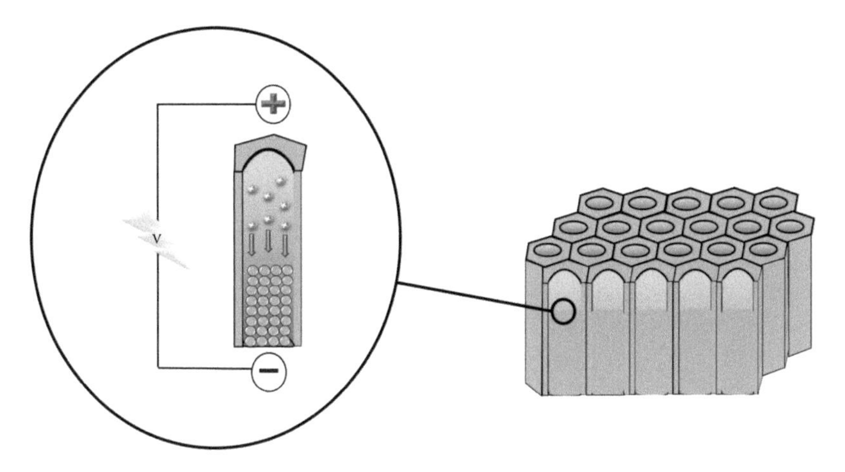

FIGURE 6.14
Diagram of the template synthesis method using the reduction of metal ions in the pores.

6.4.2 Template Synthesis of Metal Nanowires

A commonly reported method used to synthesize silver nanowires involves a template-assisted method (Bentley et al. 2005). This approach utilizes existing cylindrical channels in a template to guide the growth of nanowires (Dinh et al. 2013). Generally, nanoporous alumina or porous polycarbonate membranes are used as templates to direct the growth of nanowires (Bentley et al. 2005). The pores of the templates are filled with one or more metals via a chemical or electrochemical reduction method (Figure 6.14) (Bentley et al. 2005). After nanowire synthesis, the porous template is chemically removed, leaving metallic nanowires behind (Bentley et al. 2005).

6.4.3 Synthesis of Silver Nanowires Using a Single-Replacement Reactions

Synthesis of silver nanowires using a single-replacement reaction involves a reaction between dilute silver nitrate and a copper substrate. The substrate is neither a porous template nor a soft template. In the single-replacement reaction between the copper metal and the silver nitrate, the copper metal loses electrons and goes into the solution as ions; the silver ions in the solution then accept the electrons and deposit on the surface of the metal. For this method to work effectively, the copper substrate needs to be roughened to produce a larger number of active sites (such as surface defects), which expedite the reduction of silver ions and also serve as nucleation sites for the growth of silver nanowires (Jiang et al. 2003). Silver ions migrate to the tops of the surface defects facilitating the growth of silver wires (Figure 6.15a,b). The growth mechanism continues as silver ions diffuse to the top of the silver metal that has been reduced at the tip of the defect site. A 1D wire forms in this process, the wires do not grow laterally (Jiang et al. 2003).

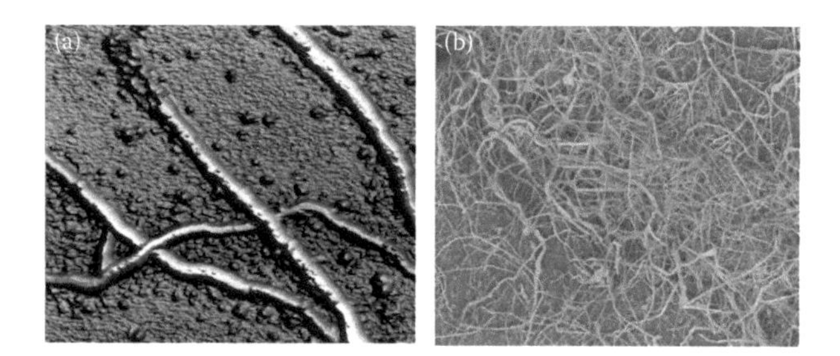

FIGURE 6.15
Atomic force microscope (a) and scanning electron microscope (b) images of silver nanowires.

6.5 Semiconductor Nanowires

Semiconductor nanowires (NWs) have exceptional electronic and optical properties, which make them ideal candidates for next-generation nanoelectronic and photonic devices (Guichard et al. 2006).

6.5.1 Semiconductor NW Synthesis

Vapor-phase synthesis is the most extensively utilized approach to semiconductor NW synthesis (Law, Goldberger and Yang 2004). A vapor phase synthesis initially involves gaseous reactants for wire formation. Among all vapor-based methods, the vapor-liquid-solid (VLS) mechanism generates large quantities of NWs with single crystalline structures. The VLS process is illustrated in Figure 6.16. VLS starts with the dissolution of gaseous

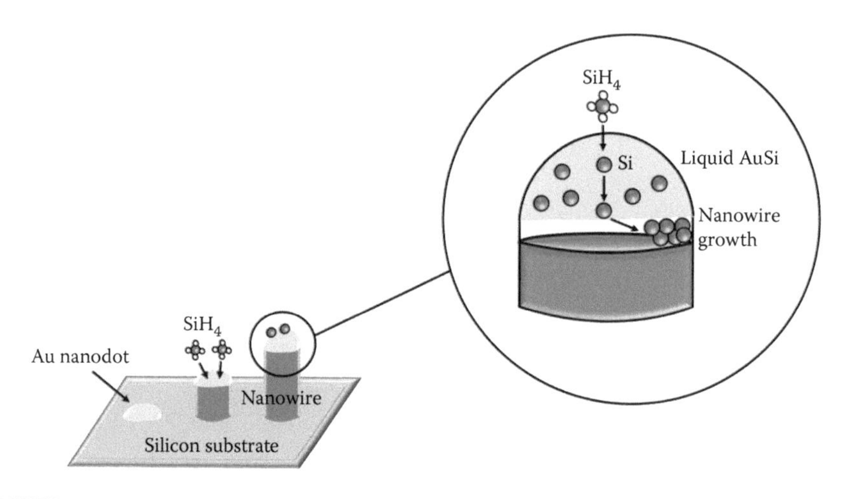

FIGURE 6.16
Vapor-liquid-solid synthesis of semiconductor nanowires.

reactants into nanosized liquid droplets of a catalyst metal. Next, nucleation and growth of single-crystalline rods occurs, followed by wire formation. The 1D growth is induced and dictated by the liquid droplets, whose sizes remain essentially unchanged during the entire process of wire growth. Each liquid droplet serves as a virtual template to strictly limit the lateral growth of an individual wire. The diameter of a NW is determined by the size of the alloy droplet, which is in turn determined by the size of the starting metal seed (Yang 2005).

6.5.2 Semiconductor Nanowire Structure

Coaxial NW (Figure 6.17) are an important class of heterostructured NWs, which are fundamentally interesting and have significant technological potential. Coaxial structures can be fabricated by coating an array of NWs with a conformal layer of a second material. The coating method chosen should allow excellent uniformity and control of the sheath thickness (Yang 2005). Cladding NWs with amorphous layers of SiO_2 or carbon is synthetically facile and has been routinely demonstrated in the literature (Yang 2005).

6.5.3 Band Gap Structure of Semiconductor NWs

The band gaps of the NWs are larger than the gap of silicon because of confinement. For a given diameter the NW band gap depends strongly on the wire type. The band-gap character also strongly depends on the crystallographic direction of the wire axis (Scheel, Reich and Thomsen 2005).

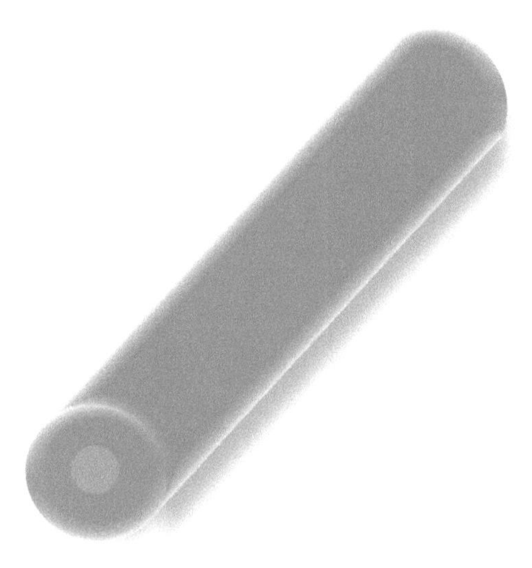

FIGURE 6.17
Coaxial heterostructured nanowire (COHN).

As the wire diameter decreases, the band gap of the NW widens and deviates gradually from that of bulk silicon. Moreover, the orientations of the wire axis and the surface have a great effect on the electronic properties of silicon NWs (Hasan, Huq and Mahmood 2013). For example, when the diameter of silicon NWs is reduced from 3.2 to 1.6 nm, the bandgap of the NWs increases from 1.56 to 2.44 eV (Hasan, Huq and Mahmood 2013).

6.5.4 Semiconductor NWs Applications

NWs are already used extensively as the key elements in nanoscale electronic devices, such as p-n junctions, field effect transistors, and logic circuits. More recently, they have also proven useful as basic building blocks for photonic components, including optically and electrically pumped lasers (Guichard et al. 2006). For example, the basic structure consisting of a single-crystalline and dopant-free NW core of germanium capped with an ultrathin epitaxial silicon shell has proven to be a remarkable system for both conventional and quantum electronics (Lieber 2011). NWs can be used to produce coherent light on the nanoscale (Yang 2005). Room-temperature UV lasing has been demonstrated with ZnO and GaN NWs. ZnO and GaN are wideband gap semiconductors with bandgaps of 3.37 and 3.42 eV, respectively, and are suitable for optoelectronic applications (Yang 2005). The possibility of controlling the band gap width is tremendously attractive for optoelectronics applications (Hasan, Huq and Mahmood 2013). It is also interesting to point out that visible and near-infrared photoluminescence (PL) at room temperature is reported from Si NWs 20 nm in diameter. The emission blue-shifted with decreasing NW diameter (Guichard et al. 2006).

6.6 End-of-Chapter Questions

1. Over _____ of nanoparticle atoms are on the surface.

   ___ 10%

   ___ 30%

   ___ 60%

   ___ 90%

2. The band gap of a metal nanoparticle _____ as the diameter of the metal nanoparticle _____.

   ___ increases, increases

   ___ increases, decreases

   ___ decreases, decreases

   ___ there is no change in band gap

3. As the particle size decreases, electrons become _____ confined in the nanoparticle.

   ___ more

   ___ less

4. Quantum confinement is achieved when electrons are _____ the Fermi Level.

   ___ above

   ___ below

5. When a metal nanoparticle is exposed to light, the _____ field component of an electromagnetic wave induces a collective, coherent oscillation of electrons in the nanoparticle.

   ___ magnetic

   ___ electric

6. The collective oscillation of electrons in a metal nanoparticle due exposure to electromagnetic waves is referred to as:

   ___ quantum confinement

   ___ surface-to-volume ratio

   ___ Fermi energy

   ___ SPR

7. The restoring force that compensates for the displacement of electrons is provided by:

   ___ magnetic fields

   ___ core electrons

   ___ nuclei

   ___ none of the above

8. _____nanoparticles absorb light with a wavelength of approximately 520 nm, and _____ nanoparticles absorb light with a wavelength of approximately 400 nm.

   ___ gold, silver

   ___ silver, gold

9. The optical properties of spherical gold nanoparticles are highly dependent on the nanoparticle diameter.

   ___ true

   ___ false

10. As nanoparticle diameter increases, _____ wavelengths are observed by metal nanoparticles.

    ___ longer

    ___ shorter

11. Gold ions (Au^{3+}) experience _____ during the Turkevich method of gold nanoparticle synthesis.
    ___ oxidation
    ___ reduction
12. Which of the following is used as a reducing agent in the Turkevich method of metal nanoparticle synthesis?
    ___ sodium chloroaurate
    ___ gold chloride
    ___ silver nitrate
    ___ sodium citrate
13. Quantum dots have physical dimensions _____ than the Bohr-exciton radius.
    ___ larger
    ___ smaller
14. Which of the following components can be used to modify the band gap of quantum dots?
    ___ core
    ___ inner semiconductor material
    ___ shell
15. Which of the following quantum dot components are made from semiconductor materials?
    ___ core
    ___ inner shell semiconductor material
    ___ shell
16. The band gap of a quantum dot _____ when the Bohr-exciton radius _____.
    ___ decreases, decreases
    ___ increases, increases
    ___ increases, decreases
    ___ there is no change in the band gap
17. When quantum dots are exposed to _____ waves with a higher energy than that of the band gap, an exciton (electron–hole pair) is formed.
    ___ infrared
    ___ microwave
    ___ visible light
    ___ ultraviolet

18. Electron–hole pairs are formed when a _____ band electron is promoted to the _____ band.

    ___ valence, conduction

    ___ conduction, valence

19. When the electron returns to a _____ energy level, emission occurs.

    ___ lower

    ___ higher

20. The emission wavelength (color) of quantum dots depends on dot size.

    ___ true

    ___ false

21. Quantum dots can emit visible light.

    ___ true

    ___ false

22. A single electron transistor (SET) is in the ON mode when the proper gate voltage lowers the energy of the _____, encouraging a single electron to tunnel through the device.

    ____ source electrode

    ____ drain electrode

    ____ quantum dot

23. Quantum dots emitting blue light have _____ band gaps and _____ diameters than quantum dots that emit red light.

    ____ smaller, larger

    ____ larger, smaller

24. In the template synthesis of metal NWs, metal ions are _____ inside of nanoscale pores of an anodic alumina oxide membrane.

    ___ oxidized

    ___ reduced

25. The band gap of a semiconductor NW _____ when the diameter _____.

    ___ increases, decreases

    ___ decreases, increases

References

Azzazy, H. M. E., M. M. H. Mansour, and S. C. Kazmierczak. 2007. "From diagnostics to therapy: Prospects of quantum dots." *Clin. Biochem.* 40: 917–927.

Bentley, A. K., M. Farhoud, A. B. Ellis, G. C. Lisensky, A. L. Nickel, and W. C. Crone. 2005. "Template Synthesis and Magnetic Manipulation of Nickel Nanowires." *J. Chem. Educ.* 82: 765–768.

Brown, T. L., H. E. LeMay, Jr., B. E. Bursten, C. J. Murphy, and P. M. Woodward. 2012. *Chemistry - The Central Science*. Boston: Prentice Hall.

Dinh, D. A., K. N. Hui, K. S. Hui, P. Kumar, and J. Singh. 2013. "Silver nanowires: A promising transparent conducting electrode material for optoelectronic and electronic applications." *Rev. Adv. Sci. Eng.* 2: 1–22.

Drbohlavova, J., V. Adam, R. Kizek, and J. Hubalek. 2009. "Quantum dots — characterization, preparation and usage in biological systems." *Int. J. Mol. Sci.* 10: 656–673.

Eustis, S., and M. A. El-Sayed. 2006. "Why gold nanoparticles are more precious than pretty gold: Noble metal surface plasmon resonance and its enhancement of the radiative and nonradiative properties of nanocrystals of different shapes." *Chem. Soc. Rev.* 35: 209–217.

Groep, J., D. Gupta, M. A. Verschuuren, M. M. Wienk, R. A. J. Janssen, and A. Polman. 2015. "Large-area soft-imprinted nanowire networks as light trapping transparent conductors." *Sci. Rep.* 5: 11414.

Guichard, A. R., D. N. Barsic, S. Sharma, T. I. Kamins, and M. L. Brongersma. 2006. "Tunable light emission from quantum-confined excitons in $TiSi_2$-catalyzed silicon nanowires." *Nano Lett.* 6: 2140–2144.

Gupta, M. 2013. "A study of single electron transistor (SET)." *Int. J. Sci. Res.* 5: 474–479.

Hasan, M., M. F. Huq, and Z. H. Mahmood. 2013. "A review on electronic and optical properties of silicon nanowire and its different growth techniques." *SpringerPlus* 2: 1–9.

Horikoshi, S., and N. Serpone. 2013. "Introduction to Nanoparticles." In *Microwaves in Nanoparticle Synthesis Fundamentals and Applications*, edited by S. Horikoshi and N. Serpone, 1–23. Weinheim: Wiley-VCH.

Huang, X., and M. A. El-Sayed. 2010. "Gold nanoparticles: optical properties and implementations in cancer diagnosis and photothermal therapy." *J. Adv. Res.* 1: 13–28.

Jain, N., R. Jain, N. Thakur, B. P. Gupta, D. K. Jain, J. Banveer, and S. Jain. 2010. "Nanotechnology: A safe and effective drug delivery system." *Asian J. Pharm. Clin. Res.* 3: 159–165.

Jamieson, T., R. Bakhshi, D. Petrova, R. Pocock, M. Imani, and A. M. Seifalian. 2007. "Biological applications of quantum dots." *Biomaterials* 28: 4717–4732.

Jiang, Z., Z. Xie, S. Zhang, S. Xie, R. Huang, and L. Zheng. 2003. "Growth of silver nanowires on metal plates by conventional redox displacement." *Chem. Phys. Lett.* 374: 645–649.

Kimling, J., M. Maier, B. Okenve, V. Kotaidis, H. Ballot, and A. Plech. 2006. "Turkevich method for gold nanoparticle synthesis revisited." *J. Phys. Chem. B* 110: 15700–15707.

Klimov, V. I. 2003. "Nanocrystal quantum dots - from fundamental photophysics to multicolor lasing." *Los Alamos Sci.* 28: 214–220.

Kolwas, K., A. Derkachova, and D. Jakubczyk. 2016. "Tailoring plasmon resonances in metal nanospheres for optical diagnostics of molecules and cells." In *Nanomedicine and Tissue Engineering State of the Art and Recent Trends*, edited by R. Augustine, N. Kalarikkal, O. S. Oluwafemi, K. S. Joshy and S. Thomas, 141–182. Boca Raton: CRC Press/Taylor & Francis Group.

Kraemer, E. O., and S. T. Dexter. 1927. "The light-scattering capacity (Tyndall Effect) and colloidal behaviour of gelatin sols and gels." *J. Phys. Chem.* 31: 764–782.

Kumar, O., and M. Kaur. 2010. "Single electron transistor: Applications and problems." *Int. J. VLSI Des. Commun. Syst.* 1: 24–29.

Law, M., J. Goldberger, and P. Yang. 2004. "Semiconductor nanowires and nanotubes." *Annu. Rev. Mater. Res.* 34: 83–122.

Leng, W., P. Pati, and P. J. Vikesland. 2015. "Room temperature seed mediated growth of gold nanoparticles: Mechanistic investigations and life cycle assesment." *Environ. Sci. Nano* 2: 440–453.

Lieber, C. M. 2011. "Semiconductor nanowires: a platform for nanoscience and nanotechnology." *MRS Bulletin* 36: 1052–1063.

Liu, C., and X. Yu. 2011. "Silver nanowire-based transparent, flexible, and conductive thin film." *Nanoscale Res. Lett.* 6: 75.

Liz-Marzán, L. M. 2004. "Nanometals: Formation and Color." *Mater. Today* 7: 26–31.

Nain, R., and R. P. Chauhan. 2009. "Colloidal synthesis of silver nano particles." *Asian J. Chem.* 21: 113–116.

NanoComposix - Color Engineering. 2017. *NanoComposix - Color Engineering.* https://nanocomposix.com/pages/color-engineering.

NanoComposix - Plasmonics. 2017. *NanoComposix - Plasmonics.* https://nanocomposix.com/pages/plasmonics.

Nikalje, A. P. 2015. "Nanotechnology and its application in medicine." *Med. Chem.* 5: 81–89.

Noguez, C. 2007. "Surface plasmons on metal nanoparticles: The influence of shape and physical environment." *J. Phys. Chem. C* 111: 3806–3819.

Pitts, A. 2008. *"13.56 MHz as a cancer cure? It just might be."* February.

Pradeep, T. 2007. *Nano: The Essentials Understanding Nanoscience and Nanotechnology.* New Delhi: Tata McGraw-Hill.

Raoof, M., and S. A. Curley. 2011. "Non-invasive radiofrequency-induced targeted hyperthermia for the treatment of hepatocellular carcinoma." *Int. J. Hepatol.* 2011: 676957.

Reed, M. A. 1993. "Quantum Dots." *Scientific American,* January. 268: 118–123.

Rogers, B., J. Adams, and S. Pennathur. 2013. *Nanotechnology The Whole Story.* Boca Raton: CRC Press/Taylor and Francis Group.

Scheel, H., S. Reich, and C. Thomsen. 2005. "Electronic band structure of high-index silicon nanowires." *Phys. Stat. Sol. B* 242: 2474–2479.

Silva, J., A. R. Fernandes, and P. V. Baptista. 2014. "Application of nanotechnology in drug delivery." In *Application of Nanotechnology in Drug Delivery,* edited by A. D. Sezer, 127–154. London: InTech.

Sreeprasad, T. S., and T. Pradeep. 2013. "Noble Metal Nanoparticles." In *Springer Handbook of Nanomaterials,* edited by R. Vajtai, 303–388. Berlin: Springer-Verlag.

Stokes, E. B., A. D. Stiff-Roberts, and C. T. Dameron. 2006. "Quantum dots in semiconductor optoelectronic devices." *Interface - Electronchemical Society* 17: 23–27.

Sun, Y., and Y. Xia. 2002. "Large-scale synthesis of uniform silver nanowires through a soft, self-seeding, polyol process." *Adv. Mater.* 14: 833–837.

Thiele, S. 2016. "Single electron transistor." In *Read-Out and Coherent Manipulation of an Isolated Nuclear Spin Using a Single-Molecule Magnet Spin Transistor,* edited by S. Thiele, 13–21. Switzerland: Springer International Publishing.

Thompson, D. 2007. "Michael Faraday's recognition of ruby gold: The birth of modern nanotechnology." *Gold Bull.* 40: 267–269.
Turkevich, J. 1985. "Colloidal gold part I. Historical and preparative aspects, morphology and structure." *Gold Bull.* 18: 86–91.
Yang, P. 2005. "The chemistry and physics of semiconductor nanowires." *MRS Bull.* 30: 85–91.
Yang, S. 2013. "Microscale synthesis and characterization of gold nanoparticles for the laboratory instruction." *Chem. Educ. J.* 15: 1–11.
Zhang, D., L. Qi, J. Ma, and H. Cheng. 2001. "Formation of silver nanowires in aqueous solutions of a double-hydrophilic block copolymer." *Chem. Mater.* 13: 2753–2755.
Zhu, J. J., C. X. Kan, J. G. Wan, M. Han, and G. H. Wang. 2011. "High-yield synthesis of Uniform Ag nanowires with high aspect ratios by Introducing the long-chain PVP in an improved polyol process." *J. Nanomater.* 2011: 1–7.

7

Nanoscale Characterization

Key Objectives

- Understand the basic operation of the following scanning probe microscopes:
  - Scanning tunneling microscope (STM)
  - Atomic force microscope (AFM)
- Become familiar with the key components of scanning probe microscopes
- Understand the operation of the feedback loop
- Understand the basic operation of the scanning electron microscope
- Become familiar with the signals produced in the scanning electron microscope, and the information provided by each signal
- Understand the basic operation of a transmission electron microscope

7.1 Introduction

Methods used to visualize and manipulate nanomaterials have been a significant factor in the emergence of nanoscience and nanotechnology (Love et al. 2005). Normal light microscopes are limited by the fact that they cannot see anything much smaller than the wavelengths of visible light. The smallest features observed with light microscopes are dependent upon the smallest wavelength of visible light utilized by the lenses. Two terms used to describe the power of a microscope are resolution and resolving power. Resolution refers to the smallest distinguishable distance between two objects (Rogers, Adams and Pennathur 2013). The human eye can see as small as 0.07 mm (Pradeep 2007). The reported resolution limit of optical microscopes is 200 nanometers (Rogers, Adams and Pennathur 2013). However, the resolving power of a microscope involves the best resolution achieved under optimum conditions; this is a property inherent to the instrument used (Pradeep 2007). The Abbe criterion describes the smallest resolvable distance between closely spaced objects (Pradeep 2007). There are several forms of microscopy available for the study of nanoscale matter, including specialized forms of optical microscopies, electron microscopies, and scanning probe microscopies (Pradeep 2007). This chapter will describe scanning probe and electron microscopes frequently used to characterize nanomaterials with at least one dimension between 1 and 100 nanometers. Scanning probe microscopy is a collective term encompassing technologies such as scanning tunneling microscopy and atomic force microscopy (Wilson et al. 2002).

7.2 Scanning Tunneling Microscopy

In the mid-twentieth century, the possibility to image or to see an atom was a matter of great debate (Rogers, Adams and Pennathur 2013). However, the STM was the first instrument to generate real-time images of surface features with an atomic resolution (Wilson et al. 2002). The STM was invented in the early 1980s by Gerd Binnig and Heinrich Rohrer of IBM; they were awarded the Nobel prize in 1986 for their invention of the STM (Rogers, Adams and Pennathur 2013).

7.2.1 Tunneling

STMs have a metal tip which scans back and forth across the surface of a sample while never touching it (Rogers, Adams and Pennathur 2013). During imaging the tip is approximately 1 nm from the sample's surface. STM operation is based on a quantum mechanical phenomenon known as "tunneling" (Hla 2005). When a voltage is applied between the tip and sample, electrons can tunnel though the gap (Hla 2005). The space between the tip and sample is an electrical barrier. However, due to the small distance between the tip and sample, the wave nature of electrons allows them to hit the barrier and reappear on the other side (Figure 7.1) (Rogers, Adams and Pennathur 2013). Miniscule changes in the distance—less than a fraction of the atomic length—can thus be detected (Hla 2005). Bringing a conductive tip and a conductive sample in very close proximity, while a small bias voltage is being applied, makes it possible for tunneling to occur (Pradeep 2007). The movement of Fermi Level electrons travelling between the tip and sample via quantum mechanical tunneling produces a measurable current and occurs among a few atoms (Pradeep 2007). The sample must be conductive; STMs in principle cannot image insulating materials (Wilson et al. 2002). By monitoring the tunneling current, the STM can track a surface's topography with picometer (10^{-12}) resolution (Rogers, Adams and Pennathur 2013). The measured tunneling current is very sensitive to the distance between the tip and the surface. Small tip-sample separation results in higher current (Baird and Shew 2004).

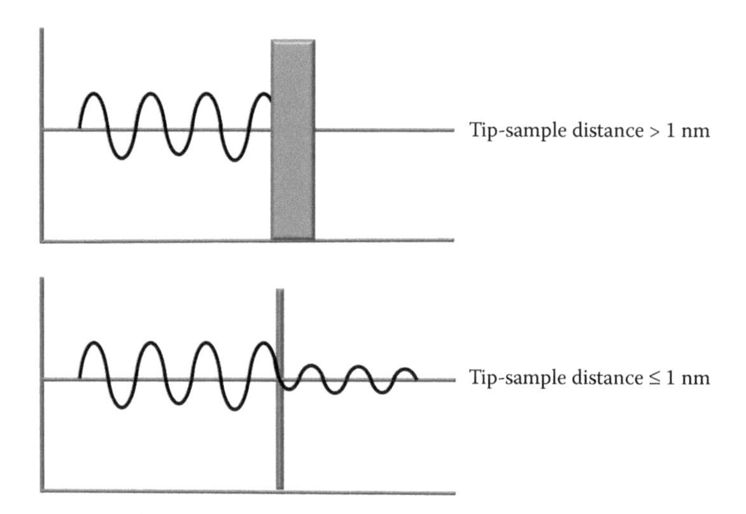

FIGURE 7.1
Effect to tip-sample distance on tunneling between the tip and sample.

7.2.2 STM Operation

A probe tip, usually made of tungsten (W) or platinum-iridium (Pt–Ir) alloy, is attached to a scanner consisting of three piezoelectric transducers, one each for x, y, and z tip motion. Upon applying a voltage, a piezoelectric transducer expands or contracts. Applying voltages to the x and y piezo allows the tip to scan the x-y plane (Chen 2008). The horizontal resolution of the STM is 200 pm. The vertical resolution of the STM can be good as 1 pm or 1/100th of an atom's diameter (Rogers, Adams and Pennathur 2013). The result is an image of the electron cloud of the surface atoms. Additionally, STM allows the study of layer growth in the semiconductor industry (Malsch 2002). STM tips are able to move atoms around, which is accomplished by varying tip-sample distance and using a voltage bias to pick up and carry, or drag a surface atom to a new position (Rogers, Adams and Pennathur 2013).

7.2.3 Feedback Loop

In constant-current mode, STMs use feedback to keep the tunneling current constant by adjusting the height of the scanner at each measurement point (Wilson et al. 2002). The feedback loop moves the tip up or down to keep the tunneling current at a constant value (Figure 7.2) (Baird and Shew 2004). If the current starts to go down, the needle is moved towards the atoms to bring the current back up, and vice versa. Recording tip distance moved up or down allows a topographic image of the surface to be produced with the aid of computer imaging software (Figure 7.3) (Baird and Shew 2004).

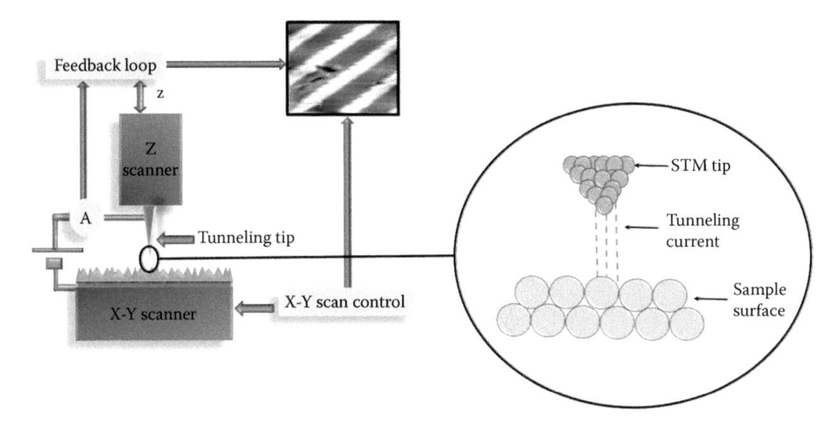

FIGURE 7.2
Operation of the STM feedback loop.

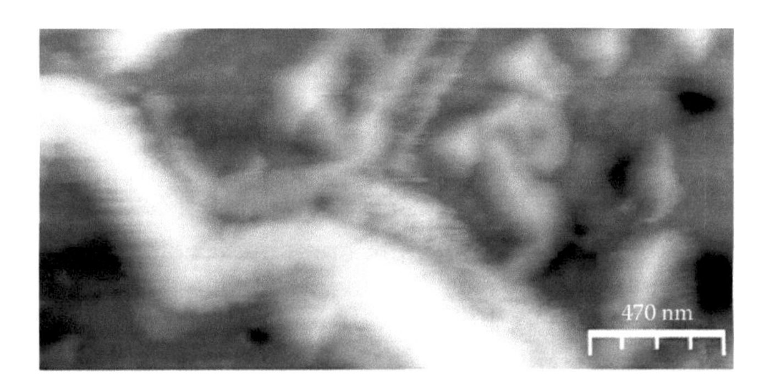

FIGURE 7.3
STM image of silver nanowire.

7.3 AFM

The AFM was invented in 1986 (Malsch 2002). Unlike STM, AFMs can investigate the surfaces of conductors and insulators (Rogers, Adams and Pennathur 2013). The basis of AFM, as a microscopic technique, is it measures the topography of the sample by generating maps of height measurements (Eaton and West 2010).

7.3.1 AFM Components

AFMs consist of a piezoelectric scanner moving a tip across the surface of a sample. A detector is used to monitor tip–sample interaction (Figure 7.4). A control station, which includes a computer and an AFM controller–electronics unit, controls the AFM's operation and generates digital images (Malsch 2002).

AFMs measure atomic forces on a surface by scanning a sharp tip attached to a flexible cantilever across the sample (Figure 7.5a,b) (Malsch 2002). AFM

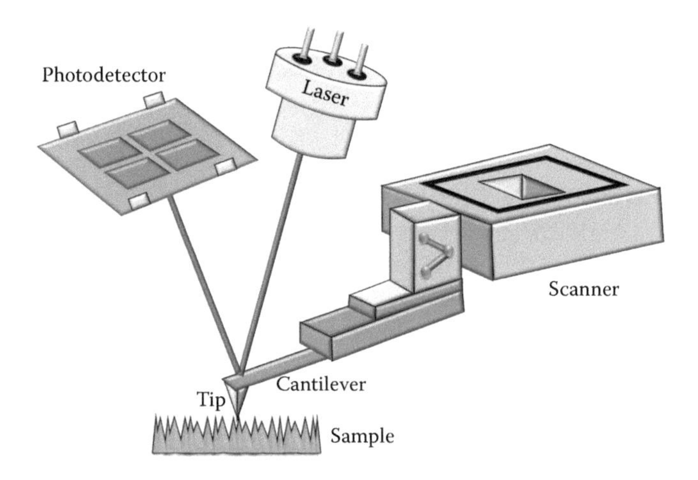

FIGURE 7.4
Diagram of AFM operation.

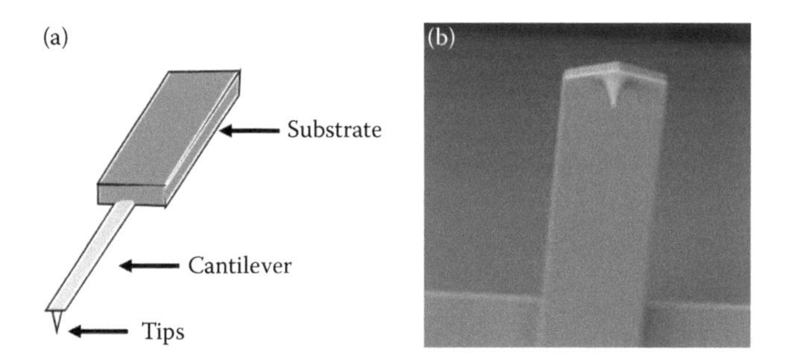

FIGURE 7.5
Diagram of AFM probe (a). SEM image of AFM probe (b).

tips are sharp enough and force sensitivity is good enough to feel forces between atoms (Rogers, Adams and Pennathur 2013). The tip diameter can be as small as 5 nm (Malsch 2002).

AFM tips are moved across a sample surface using a scanner containing piezoelectric materials. Piezoelectric materials convert electrical signals into mechanical motion. Typically, the expansion coefficient for a single piezoelectric device is on the order of 0.1 nm per applied volt. This means if the voltage applied to the piezoelectric device is 2 volts, then the material will expand approximately 0.2 nm, which is approximately the diameter of a single atom (Malsch 2002).

7.3.2 AFM Operation

As the tip approaches the surface, attractive intermolecular forces exist between surface and tip atoms. When the cantilever is pushed into the surface, repulsive forces are observed because the tip is attempting to displace the atoms on the surface. Attractive and repulsive forces displace the cantilever (deflection); and cantilever displacement is used to measure surface topography and properties (Wilson et al. 2002). The AFM enables the measurement of forces as small as 10^{-12} N (Malsch 2002). The sample can be imaged in air, liquid, or vacuum (Rogers, Adams and Pennathur 2013). As an AFM tip tracks the sample surface, the force between the tip and the surface causes the cantilever to bend. An optical sensor measures the deflection of the cantilever. The most common optical sensor consists of a laser beam reflected from the metal-coated back of the cantilever onto a positional sensitive photodiode. The positional sensitive photodiode can measure changes in position of the incident laser beam as small as 1 nm, thus enabling subnanometer resolution (Wilson et al. 2002). The piezoelectric scanner moves the tip over the sample surface. The force sensor monitors the force between the tip and the surface, and the feedback control feeds the signal from the force sensor back into the piezoelectric to maintain a fixed force between the tip and sample (Malsch 2002). The feedback loop (Figure 7.6) is used to maintain a set force between the probe and the sample. If there is an increase in force (when the tip encounters a raised region on the sample), the feedback control causes the piezoelectric scanner to move the probe away from the surface. Conversely, if the optical sensor detects a decrease in force, the probe is moved towards the surface. The amount the z piezoelectric moves up and down to maintain the tip–sample distance is assumed to be equal to the sample topography. In this way, by monitoring the voltage applied to the z piezo, a map of the surface shape (height image) is measured. The

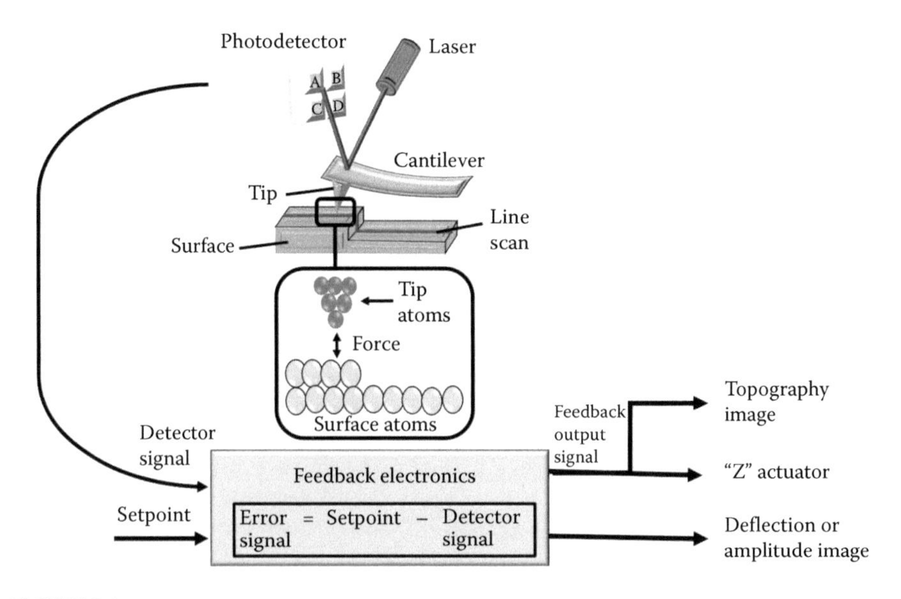

FIGURE 7.6
Schematic of the operation of an AFM feedback loop.

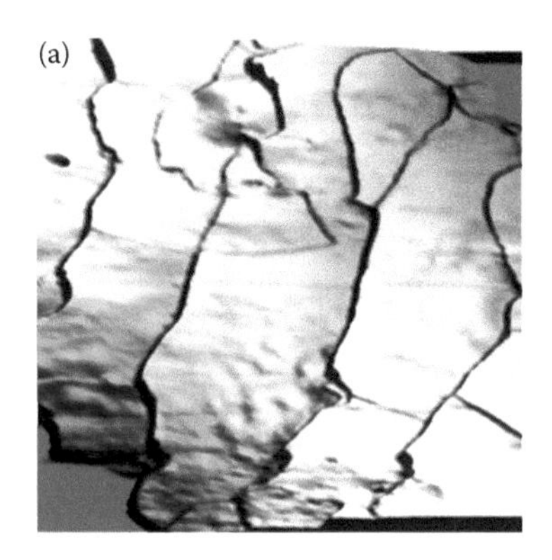

FIGURE 7.7
3D AFM image of human hair (a) and 2D AFM image of bacteria (b).

amount the z piezo moves to maintain the deflection set point is taken to be the sample topography. This signal, plotted versus distance, forms the height or topography image in contact-mode AFM (Figure 7.7) (Eaton and West 2010). This signal is used to create a 3D image of the surface structure displayed on a screen (Figure 7.7a,b) (Malsch 2002; Rogers, Adams and Pennathur 2013).

7.3.3 Operating Modes of AFM

Contact mode involves direct contact between the tip and the sample surface. When the AFM tip encounters surface features, the AFM system adjusts the vertical position of the tip so that the force between the tip and sample remains constant (Rogers, Adams and Pennathur 2013). Consequently, during contact mode, the tip makes direct contact with the surface during scanning and can cause physical damage to soft materials. In tapping mode, the cantilever is vibrated at or near its resonance frequency allowing the tip to contact the surface intermittently (Figure 7.8a) (Wilson et al. 2002). Intermittent contact reduces possible surface damage or tip contamination while maintaining resolution (Wilson et al. 2002). In noncontact mode, the attractive intermolecular forces

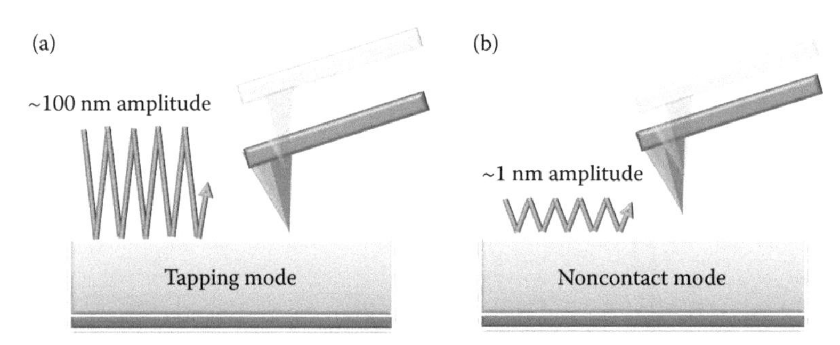

FIGURE 7.8
Tapping mode AFM (a) and noncontact mode AFM (b).

are measured by oscillating the cantilever approximately 5 to 10 nm above the sample surface (Figure 7.8b) (Wilson et al. 2002). When the oscillating probe approaches the sample surface, the oscillation changes due to the interaction between the probe and the forces from the sample. This leads to a reduction in the frequency and amplitude of the oscillation. The oscillation is monitored by the optical sensor, and the scanner adjusts the z height via the feedback loop to maintain a constant vibrational amplitude (Eaton and West 2010).

7.3.4 Force Curves

Tip-sample interactions can be used to generate a force curve (Figure 7.9). Force curve data can be used to examine forces between the tip and the sample. Regions in force curves includes the following action:

A—the cantilever is lowered to the surface

B—tip snaps into contact when it is 1–10 nm away from the sample surface

C—as the cantilever is lowered, the cantilever bends up because the tip is pushing against the sample surface

D—the cantilever is pulled away from the sample surface and the tip holds onto the surface causing the cantilever to bend down due to intermolecular interactions between the surface atoms and tip atoms

E—tip detaches from the sample surface (Rogers, Adams and Pennathur 2013).

Force curve data allows scientists to study nanomechanical phenomena, as shown in Figure 7.10. In these experiments, one end of the molecule is

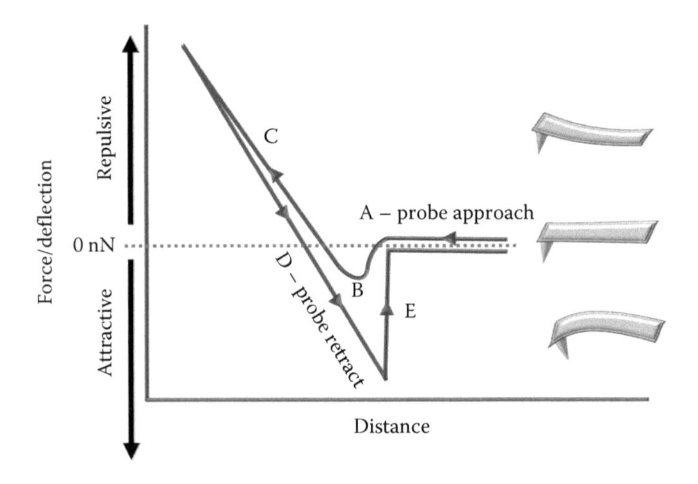

FIGURE 7.9
Example of a force curve.

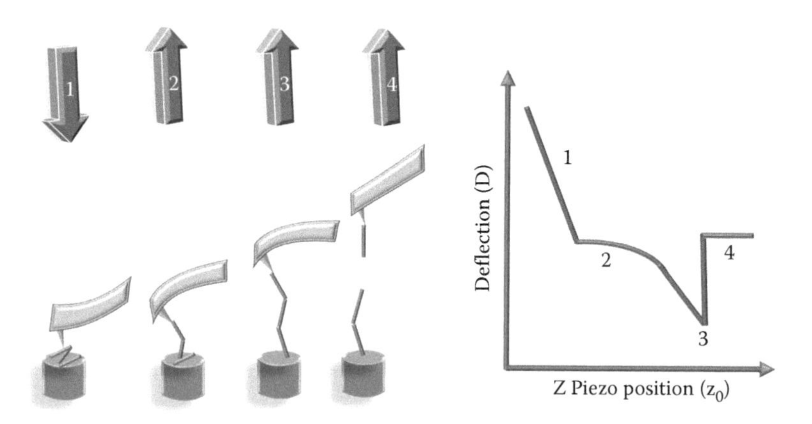

FIGURE 7.10
Force curve analysis used for the determination of the mechanical properties of molecules. Molecules are tethered between the tip and sample (1). The tip is withdrawn from the sample (2). The maximum force is applied to the molecule without breaking as the tip continues to move away from the sample surface (3). Rupture of the molecular tether (4).

attached to the AFM tip and the other end is attached to the surface of a substrate. Data collected from force curve measurements includes adhesive forces and rupture forces needed to break the bonds in a molecule (Rogers, Adams and Pennathur 2013).

7.3.5 Nanoshaving

The sharpness of tips used in atomic force microscopes has allowed researchers to move atoms on surfaces and to fabricate nanopatterns in thin films (Yang, Amro and Liu 2003). In nanoshaving (Figure 7.11a), the AFM tip exerts high force on a sample surface. This pressure causes the displacement of absorbed molecules or thin films. Holes and trenches can thus be fabricated (Figure 7.11b) (Yang, Amro and Liu 2003). This procedure allows precise control over the size and geometry of the fabricated features. Additionally, this technique makes it possible to obtain an edge resolution better than 2 nm (Xu and Liu 1997).

7.3.6 Dip-Pen Nanolithography

Dip-pen nanolithography (DPN) uses an "ink"-coated AFM tip to pattern surfaces (Figure 7.12a). DPN is a direct-write "additive" process, which allows soft and hard materials to be printed from AFM tips onto a surface substrate. It is important to point out, that no premodification of the substrate surface is required prior to DPN (Salaita, Wang and Mirkin 2007). DPN can deposit organic molecules via a meniscus onto a substrate surface under ambient conditions (Ivanisevic and Mirkin 2001). With DPN, an organic molecule

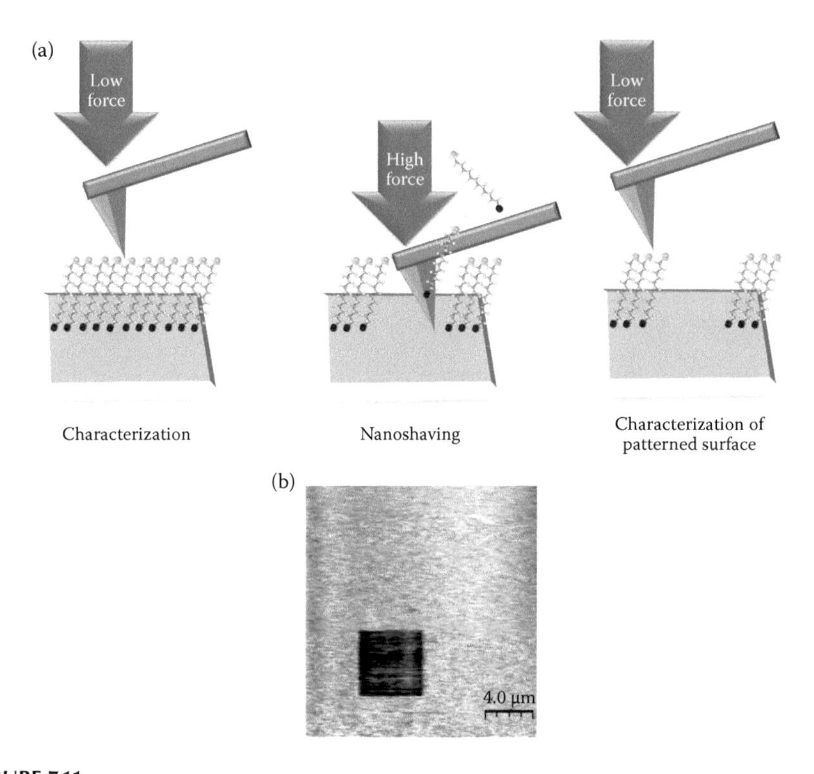

FIGURE 7.11
Nanoshaving a pattern on a SAM surface (a). A patterned etched in a mercaptohexadecanoic acid self-assembled monolayer (SAM) (b).

such as a thiol is used as "ink" and is transported from an AFM tip through a liquid meniscus onto a gold surface (Ivanisevic and Mirkin 2001). One-molecule thick nanostructures with micron to sub-100 nm dimensions can be fabricated with DPN (Figure 7.12b) (Ivanisevic and Mirkin 2001).

7.4 Scanning Electron Microscopy

Electron microscopy is a technique involving the use of an electron beam to form magnified images of specimens. The principle advantage of using electrons, rather than light, to form magnified images is electrons provide as much as a thousand-fold increase in resolving power. The resolving power of a modern light microscope is approximately 200 nm. A transmission electron microscope (TEM) can resolve detail to approximately 0.2 nm, and a scanning electron microscope (SEM) to approximately 3 nm (Flegler, Heckman and Klomparens 1993). Development of the SEM began in 1935 with Max Knoll at the Technical University in Berlin. Knoll used an electron beam to scan

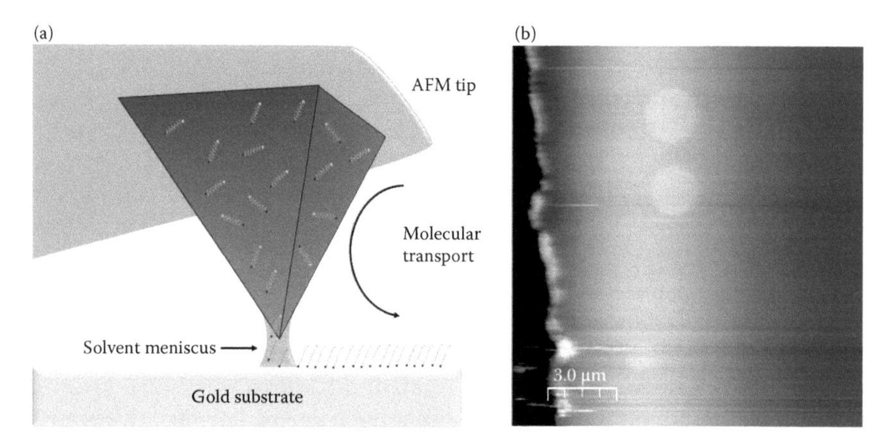

FIGURE 7.12
Process of dip-pen nanolithography (a). Alkanethiols deposited on a gold surface using dip-pen nanolithography (b).

specimens in a modified cathode ray-tube. However, the instrument could not see samples at high magnification due to the use of an electron beam with a large diameter. Electrostatic lenses were used in later electron microscopes to produce a narrower electron beam with a diameter of 50 nm. Further improvements followed when electromagnetic, rather than electrostatic lenses were used; this produced electron beams with diameters less than 20 nm (Wilson et al. 2002). In current electron microscopes (Figure 7.13), the lenses used are electromagnetic

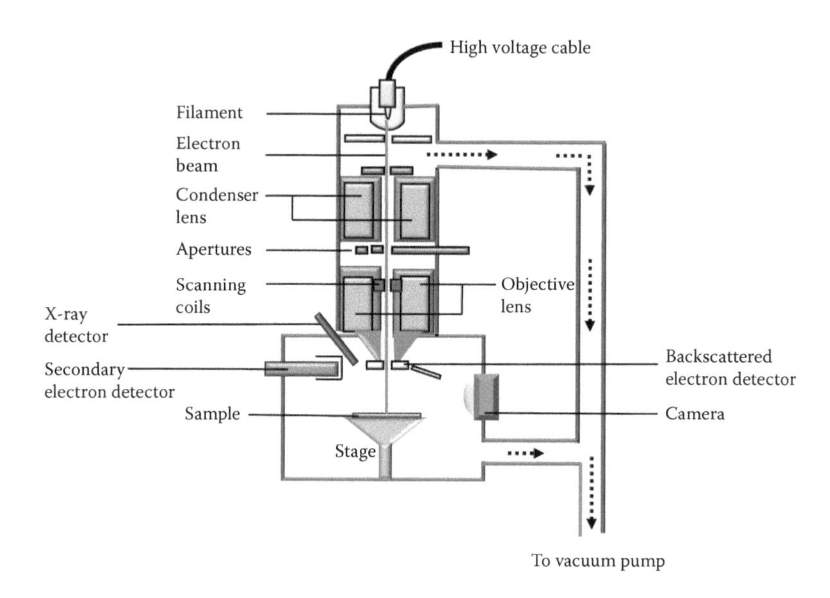

FIGURE 7.13
Schematic diagram of a SEM.

in nature (Flegler, Heckman and Klomparens 1993). In a SEM, an electron gun emits a beam of electrons, which passes through a condenser lens and is refined into a thin stream (Wilson et al. 2002). A condenser lens condenses the beam of electrons using electromagnetic coils (Flegler, Heckman and Klomparens 1993). SEM is primarily used to image surface morphology. Since electron microscopes use electrons instead of light to carry the information generated during image formation, all images are black and white. Color itself is a function of visible light, and no visible light is used to generate images in an electron microscope (Flegler, Heckman and Klomparens 1993).

The SEM has a large depth of field (the amount of the sample that can be in sharp focus at one time). Depth of field can be up to four hundred times greater than a light microscope (Flegler, Heckman and Klomparens 1993). Apertures are used in conjunction with the electromagnetic lenses. An aperture is a piece of metal with a small round hole to limit the diameter of the beam of electrons. They are also used to eliminate stray or widely scattered electrons (Flegler, Heckman and Klomparens 1993). Vacuum systems are a necessity in the operation of all electron microscopes. An electron beam cannot be generated or maintained in a gas-filled environment because gas within the chamber (oxygen) and the heated electron emitter react, and the filament burns out. At room temperature and pressure, a cubic meter of air contains about 2.5×10^{25} molecules with a mean free path of about 65 nm between collisions. A typical vacuum reached in an electron microscope column contains gases with a mean free path of 6.5 m (Flegler, Heckman and Klomparens 1993).

7.4.1 Lenses and Scan Coils

The electron beam in a SEM is controlled using a series of electromagnets (Pradeep 2007). Electromagnetic lenses consist of coils of wire enclosed in an iron casing and are used to create a magnetic field upon application of current in the coils (Pradeep 2007). Coils of wire, called the scan coils, are located within the objective lens. A varying voltage is applied to the scan coils, creating a magnetic field which deflects the beam of electrons back and forth in a pattern called a raster (Flegler, Heckman and Klomparens 1993). Scan coils allows images to be formed using a point-by-point method. A set of scan coils for the x-direction and the y-direction can be found in scanning electron microscopes (Pradeep 2007). The scan coils move the electron beam across the specimen (Pradeep 2007).

7.4.2 Electron Guns

Electrons are generated in the electron gun, located at the top of the column. The major types of electrons guns used in electron microscopes are tungsten filaments, lanthanum-hexaboride guns, and field emission guns. Tungsten filaments are the most common (Flegler, Heckman and Klomparens 1993). A tungsten filament is referred to as a thermionic emitter because electrons are emitted by exceeding the work function of the metal by using a high surface temperature between 2000 and 2700 K (Pradeep 2007). There are some disadvantages associated with tungsten-filament electron guns. Filaments evaporate as a result of high operating temperatures, and as a result, the wire gets thin enough to break (Pradeep 2007). Lanthanum-hexaboride (LaB_6) guns contain a crystal of LaB_6 that is heated and emits electrons. These electron guns are brighter than tungsten guns and produce a smaller electron beam diameter (Flegler, Heckman and Klomparens 1993). Field emission guns use a very high voltage to pull electrons off the wire. These electron guns are referred to as cold cathode guns because heating is not used to generate electrons. Field emission guns are a thousand times brighter than tungsten filaments, have a smaller beam diameter, and possess higher lifetimes (Flegler, Heckman and Klomparens 1993).

7.4.3 Beam-Specimen Interactions

When incident-beam electrons strike the surface of a sample, they undergo a series of complex interactions with the nuclei of the atoms in the sample. The interaction is usually described as teardrop or pear shaped (Figure 7.14). In samples, the volume (depth and width) of the interaction varies directly with the acceleration voltage and inversely with the average atomic number of the sample (Zhou and Wang, Scanning Microscopy for Nanotechnology 2006).

7.4.3.1 Secondary Electrons

When the primary beam strikes sample surfaces, it results in ionization of specimen atoms. This encourages loosely bound electrons to be emitted (Figure 7.15). These electrons are referred to as secondary electrons. They have low energy, so they can only escape from a region within a few nanometers of the material's surface. The maximum escape depth had been calculated as 5 nm in metals and 50 nm in insulators (Zhou and Wang, Scanning Microscopy for Nanotechnology 2006). Secondary electrons are used to produce topographic images with good resolution (Figure 7.16a–c) (Zhou, Apkarian et al. 2007).

7.4.3.2 Backscattered Electrons

When an electron from the beam encounters a nucleus in the specimen, the resultant attraction produces a deflection in the electron's path (Wilson et al. 2002). Backscattered electrons are beam electrons which have been scattered backward. They are scattered out of the sample from the same side they entered (Figure 7.17) (Zhou and Wang, Scanning Microscopy for Nanotechnology 2006). A few of these electrons will be completely backscattered, reemerging from the surface of the sample (Wilson et al. 2002).

Backscattered electrons are not strongly absorbed by the sample because of their high energy, and a high percentage do escape the sample. The maximum

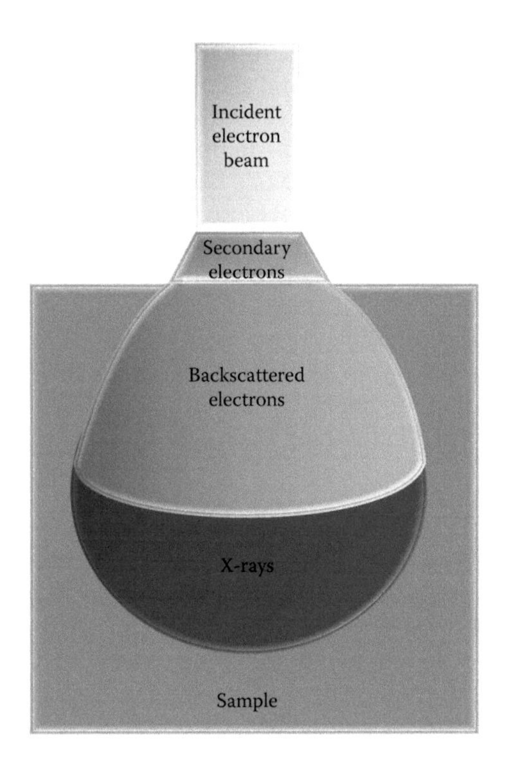

FIGURE 7.14
Interaction volume and signals generated during beam/specimen interactions.

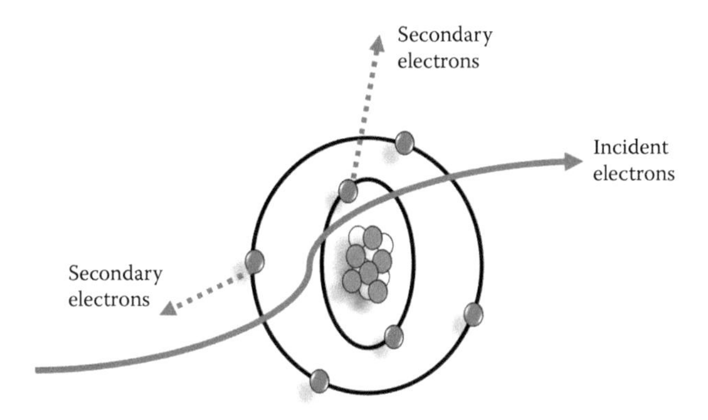

FIGURE 7.15
Production of secondary electrons.

escape depth (and width) can range from a fraction of a micrometer to several micrometers. Due to the large escape depth, the backscattered electron images are less sensitive to differences in surface topography (Zhou and Wang, Scanning Microscopy for Nanotechnology 2006). Since the extent of scattering is strongly dependent on the atomic number of the nucleus involved, the

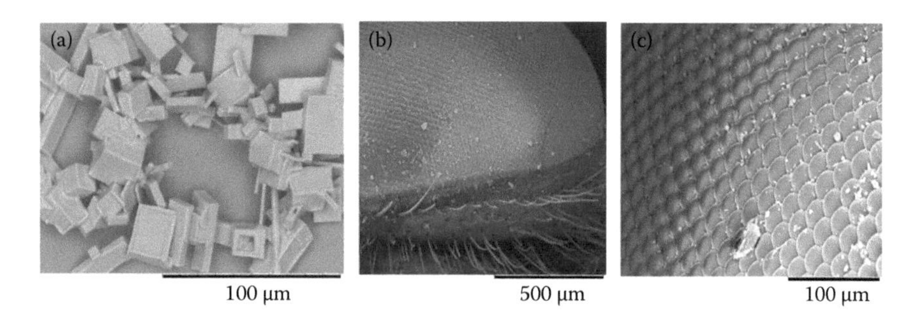

FIGURE 7.16
Secondary electron images of table salt (a), a fly eye at 150× magnification (b), and 600× magnification (c).

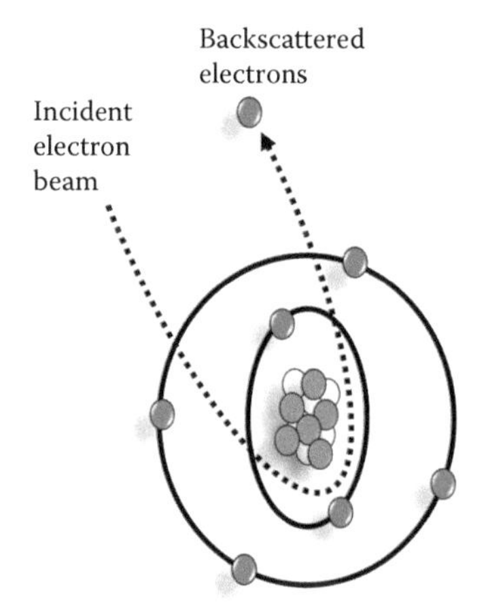

FIGURE 7.17
Production of backscattered electrons.

backscattered electrons produce images containing information about a sample's atomic composition (Wilson et al. 2002). Backscattered electrons are very useful in detecting the presence of differences in average atomic number of a sample (Zhou and Wang, Scanning Microscopy for Nanotechnology 2006). Elements with higher atomic numbers produce more backscattered electrons and appear brighter than the components with lower atomic numbers (Figure 7.18) (Zhou and Wang, Scanning Microscopy for Nanotechnology 2006).

7.4.3.3 X-Rays

The analysis of characteristic X-rays to provide chemical information is the most widely used microanalytical technique in the SEM. When an inner shell electron is displaced by collision with an incident electron, an outer shell electron may fall into the inner shell. When this happens, the resulting imbalance may

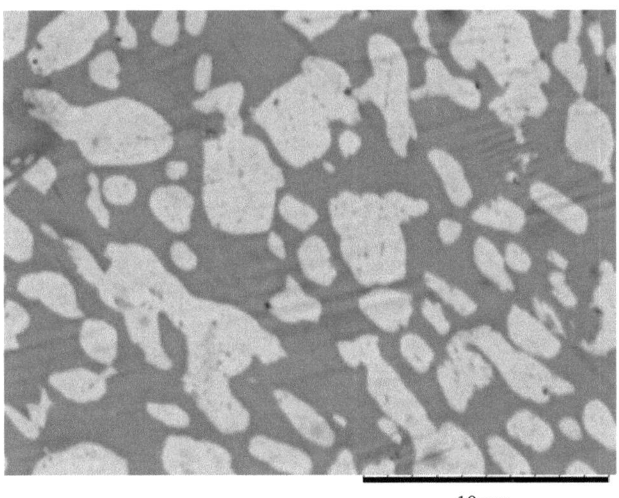

FIGURE 7.18
Backscattered electron image of lead-tin solder. The brighter areas correspond with lead and darker areas with tin.

be corrected by the production of an X-ray (Figure 7.19) (Zhou et al. 2007). This technique is referred to as energy-dispersive X-ray spectroscopy (EDS), and is commonly used for the elemental analysis or chemical characterization of a sample (Flegler, Heckman and Klomparens 1993).

Measurement of the energy or the wavelength of the resulting X-ray can be used to determine the elemental composition of the sample (Figure 7.20) (Zhou and Wang, Scanning Microscopy for Nanotechnology 2006).

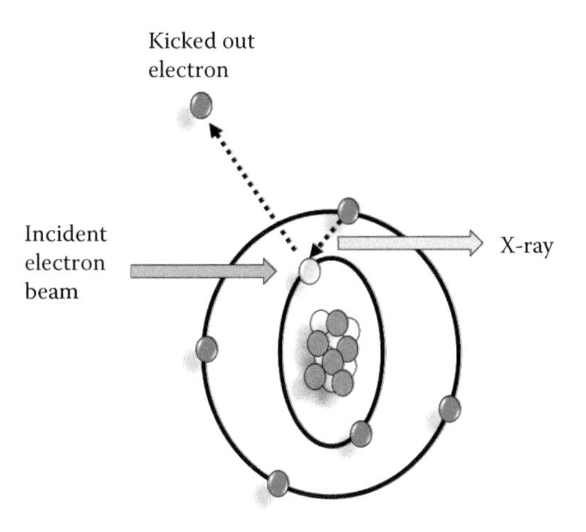

FIGURE 7.19
Production of an X-ray.

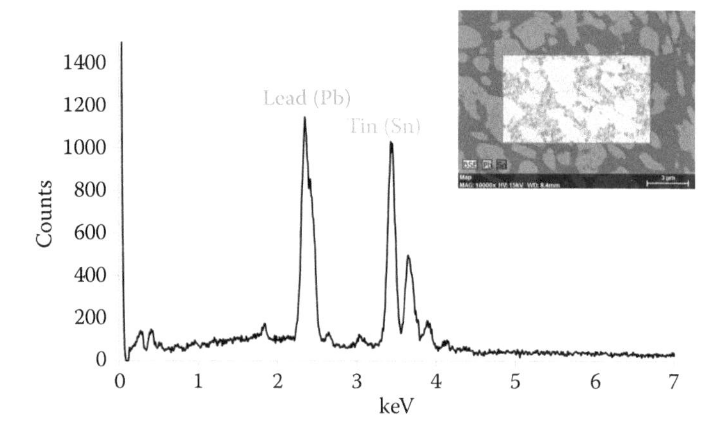

FIGURE 7.20
X-ray analysis of tin-lead solder.

7.5 TEM

In TEM, transmitted electrons are used to create magnified images of samples (Pradeep 2007). A TEM is primarily used for imaging thin sections or small particles (Flegler, Heckman and Klomparens 1993). The TEM has a resolving power of approximately 0.2 nm (Wilson et al. 2002). To produce a standard TEM image, the electron beam must be able to penetrate the sample (Wilson et al. 2002). A TEM works much like a slide projector, where a beam of light passes through the slide. The light passing through the slide is subjected to changes by structures and objects on the slide and as a result, only certain parts of the light beam is transmitted through certain parts of the slide. Material thicknesses for TEM samples are prepared in advance to allow electrons to be transmitted through the sample (Wilson et al. 2002). A scattering mechanism is used to generate an image in TEM (Pradeep 2007). Scattering occurs when the electron beam interacts with matter. The recombination of transmitted and scattered beams produces a phase contrast image (Pradeep 2007).

7.5.1 TEM Components

At the top of the TEM column (Figure 7.21) is a filament producing a stream of electrons. The stream is focused to a small, thin, coherent beam using a condenser lens (Flegler, Heckman and Klomparens 1993). Additionally, apertures protect the specimen from too many stray electrons. The condenser-lens system also controls electron illumination on the specimen (Flegler, Heckman and Klomparens 1993). The beam strikes the specimen and parts of it are transmitted. In TEM images, the darker areas represent areas of the sample through which fewer electrons were transmitted (those areas were thicker or denser in the sample); the lighter areas of the image represent

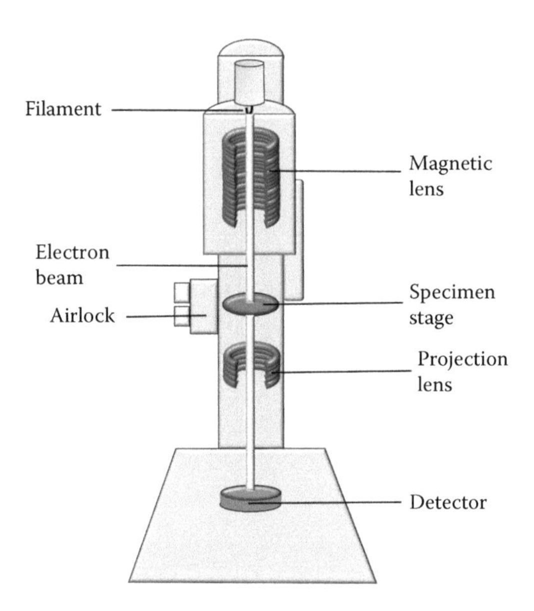

FIGURE 7.21
Components of a TEM.

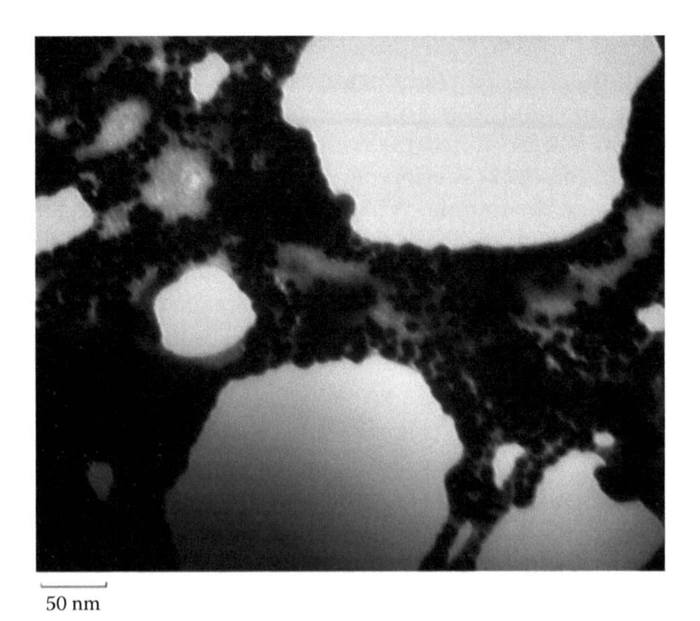

FIGURE 7.22
TEM image of gold nanoparticles.

those areas of the sample through which more electrons were transmitted (those areas where thinner or less dense), as shown in the TEM image of gold nanoparticles in Figure 7.22 (Wilson et al. 2002).

7.6 End-of-Chapter Questions

1. The smallest distinguishable distance between two objects is referred to as:
   ___ resolving power
   ___ resolution
2. Which of the following is an inherent property of the microscope used?
   ___ resolving power
   ___ resolution
3. Scanning probe microscopy is a term that describes:
   ___ scanning tunneling microscopy and scanning electron microscopy
   ___ scanning electron microscopy and transmission electron microscopy
   ___ scanning electron microscopy and atomic force microscopy
   ___ scanning tunneling microscopy and atomic force microscopy
4. Which of the following generated the first real-space images of surfaces with atomic resolution?
   ___ the AFM
   ___ the STM
   ___ the SEM
   ___ the TEM
5. The STM functions by scanning a tip along a conducting surface and measuring _____ between the tip and sample.
   ___ atomic forces
   ___ intermolecular forces
   ___ hydrogen bonding
   ___ tunneling current
6. Tunneling current _____ exponentially with decreasing tip-sample distance.
   ___ increases
   ___ decreases

7. Separation distance between the tip and sample in scanning tunneling microscopy needs to be approximately _____ nm in order to detect tunneling current.

   ___ 20

   ___ 10

   ___ 5

   ___ 1

8. Conductive and nonconductive samples can be imaged with scanning tunneling microscopy.

   ___ true

   ___ false

9. A small _____ applied between the tip and sample is necessary for electron tunneling to occur.

   ___ force

   ___ voltage

10. The horizontal resolution of the STM is _____ pm and the vertical resolution _____ pm.

    ___ 1, 200

    ___ 200, 1

    ___ 100, 20

    ___ 20, 100

11. What is used to keep the tunneling current constant during STM imaging?

    ___ tip

    ___ tip voltage

    ___ feedback loop

    ___ sample conductivity

12. The feedback loop in STM keeps tunneling current constant by adjusting:

    ___ tip height

    ___ sample conductivity

    ___ tip voltage

    ___ tunneling current cannot be kept constant

13. AFM tip diameters can be as small as:

    ___ 10 nm

    ___ 5 nm

    ___ 15 nm

    ___ 20 nm

14. The AFM scanner uses _____ materials to move a tip across a sample surface.

    ___ electromechanical

    ___ piezoelectric

    ___ glass

    ___ graphite

15. AFMs measure _____ on a surface by scanning a sharp tip attached to a flexible cantilever across the sample.

    ___ magnetism

    ___ electricity

    ___ atomic forces

    ___ light emission

16. The AFM can measure forces as small as:

    ___ 10^{-2} N

    ___ 10^{-5} N

    ___ 10^{-12} N

    ___ 10^{-7} N

17. The feedback loop in atomic force microscopy (AFM) is used to maintain constant _____ between the tip and sample.

    ___ magnetism

    ___ force

    ___ electrostatic attraction

    ___ none of the above

18. Which of the following is NOT an operating mode of AFM?

    ___ tapping mode

    ___ contact mode

    ___ scanning electron mode

    ___ noncontact mode

19. In _____, the cantilever pushes lightly against the surface and follows the surface structure.

    ___ contact mode

    ___ tapping mode

20. Which AFM mode is less disruptive to soft samples such as delicate biological molecules or certain surface features?

    ___ contact mode

    ___ tapping mode

21. Nanomechanical information can be obtained using:
    ___ topography data
    ___ force curve data
    ___ the feedback loop
    ___ none of the above
22. The form of AFM lithography involving the removal of material on a substrate surface is:
    ___ vnanoshaving
    ___ dip pen nanolithography
23. The form of AFM lithography involving the addition of material on a substrate surface is:
    ___ nanoshaving
    ___ dip pen nanolithography
24. What generates the electron beam in the scanning electron microscope?
    ___ the anode
    ___ the electron gun
    ___ the scanning coil
    ___ the electron detector
25. _____ act as lenses in a scanning electron microscope (SEM).
    ___ glass
    ___ clear plastic
    ___ electromagnetic coils
    ___ thin films
26. What is used to move or raster the electron beam across the surface of a sample in SEM?
    ___ condenser coils
    ___ scan coils
    ___ electron gun
    ___ aperture
27. Which of the following is necessary for the successful operation of a SEM?
    ___ a tip
    ___ a cantilever
    ___ a vacuum
    ___ a scanner

28. In a SEM, loosely bound electrons that are emitted from the sample are referred to as:
___ backscatter electrons
___ X-rays
___ secondary electrons
___ none of the above
29. In a SEM, electrons that are scattered out of the sample are referred to as:
___ X-rays
___ backscatter electrons
___ secondary electrons
___ none of the above
30. Which of the following signals is used to produce topographical images?
___ X-rays
___ backscattered electrons
___ secondary electrons
___ none of the above
31. Which of the following signals can be used to obtain compositional information?
___ X-rays
___ backscattered electrons
___ secondary electrons
___ none of the above
32. Which of the following signals is used for elemental analysis?
___ X-rays
___ backscattered electrons
___ secondary electrons
___ none of the above.
33. Electrons _______ are used to produce 2D images in a TEM.
___ reflected off the sample
___ transmitted through the sample
___ emitted by the sample
___ none of the above
34. TEMs are used to primarily image what type of samples?
___ thin
___ conductive

___ nonconductive

___ radioactive

35. Darker areas in TEM images represent _____ areas in a sample.

___ thicker

___ thinner

36. Brighter areas in TEM images represent _____ areas in a sample.

___ thicker

___ thinner

References

Baird, D., and A. Shew. 2004. "Probing the history of scanning tunneling microscopy." In *Discovering the Nanoscale*, edited by D. Baird, A. Nordmann and J. Schummer, 145–156. Amsterdam: IOS Press.

Chen, J. C. 2008. *Introduction to Scanning Tunneling Microscopy*. New York: Oxford University Press.

Eaton, P., and P. West. 2010. *Atomic Force Microscopy*. Oxford: Oxford University Press.

Flegler, S. L., J. W. Heckman, and K. L. Klomparens. 1993. *Scanning and Transmission Electron Microscopy: An Introduction*. New York: Oxford University Press.

Hla, S. 2005. "STM single atom/molecule manipulation and it's application to nanoscience and nanotechnology." *J. Vac. Sci. Tech.* 23: 1351–1360.

Ivanisevic, A., and C. A. Mirkin. 2001. ""Dip-pen" nanolithography on semiconductor surfaces." *J. Am. Chem. Soc.* 123: 7887–7889.

Love, J. C., L. A. Estroff, J. K. Kriebel, R. G. Nuzzo, and G. M. Whitesides. 2005. "Self-Assembled Monolayers of Thiolates on Metals as a Form of Nanotechnology." *Chem. Rev.* 105: 1103–1169.

Malsch, I. 2002. "Tiny tips probe nanotechnology." *The Industrial Physicist*, October: 16–19.

Pradeep, T. 2007. *Nano: The Essentials Understanding Nanoscience and Nanotechnology*. New Delhi: Tata McGraw-Hill.

Rogers, B., J. Adams, and S. Pennathur. 2013. *Nanotechnology – The Whole Story*. Boca Raton: CRC Press/Taylor & Francis Group.

Salaita, K., Y. Wang, and C. A. Mirkin. 2007. "Applications of dip-pen nanolithography." *Nat. Nanotechnol.* 2: 145–155.

Wilson, M., K. Kannangara, G. Smith, M. Simmons, and B. Raguse. 2002. *Nanotechnology – Basic Science and Emerging Applications*. Boca Raton: Chapman and Hall/CRC Press.

Xu, S., and G. Liu. 1997. "Nanometer-scale fabrication by simultaneous nanoshaving and molecular self-assembly." *Langmuir* 13: 127–129.

Yang, G., N. A. Amro, and G. Liu. 2003. "Scanning probe lithography of self-assembled monolayers." In *Nanofabrication Technologies*, edited by E. A. Dobisz, 52–65. Bellingham: SPIE.

Zhou, W., R. P. Apkarian, Z. L. Wang, and D. Joy. 2007. "Fundamentals of scanning electron microscopy." In *Scanning Microscopy for Nanotechnology*, edited by W. Zhou and Z. L. Wang, 1–40. New York: Springer-Verlag.

Zhou, W., and Z. L. Wang. 2006. *Scanning Microscopy for Nanotechnology*. New York: Springer.

8

Nanofabrication Techniques

Key Objectives

- Understand the basic principles associated with various forms of soft lithography
- Understand the basic principles associated with physical vapor deposition and chemical vapor deposition
- Understand the basic operation of sputter coaters and metal evaporation systems
- Know characteristics associated with wet and dry etches
- Become familiar with the basic operating principles of electron and ion-based lithography systems
- Become familiar with the major steps associated with photolithography

8.1 Introduction

There are two distinct approaches used to fabricate nanomaterials. The "top-down" approach involves taking a block of material and decreasing the size until the desired dimensions are achieved. The size limit of the smallest features produced depends on the tools used. "Bottom-up" nanofabrication utilizes molecular or atomic building blocks to fit together to create larger objects (Wilson et al. 2002). This chapter will describe commonly used methods for nanomaterial fabrication.

8.2 Soft Lithography

Soft lithography utilizes an elastomeric stamp with patterned structures on its surface to create nanostructures with feature sizes ranging from 30 nm to

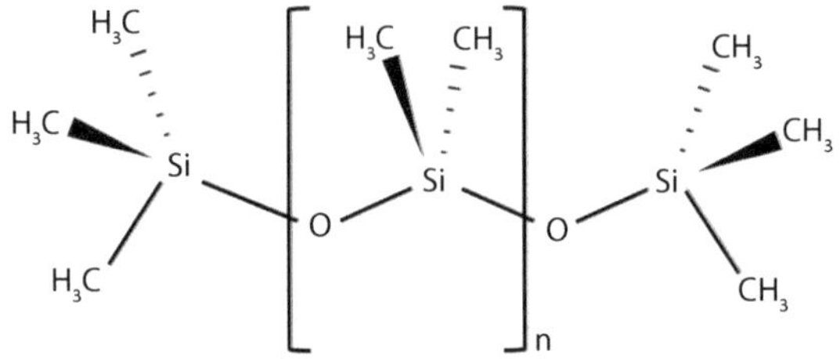

FIGURE 8.1
Structure of polydimethylsiloxane (PDMS).

FIGURE 8.2
PDMS stamp made using a compact disc hard master.

100 μm in size (Sanders 2015). Soft lithography provides a convenient, effective, and low-cost method for creating nanostructures. The stamps used for soft lithography are made from polydimethylsiloxane (PDMS) (Figure 8.1) (Lyman et al. 2010; Sanders 2015). PDMS stamps make conformal contact with surfaces and can be released easily from hard masters (Xia and Whitesides 1998). To form a stamp, the liquid elastomer is poured over a hard master possessing the patterns to be replicated. The mixture is heated to elevated temperatures, and in the process, the mixture becomes a solid (Xia and Whitesides 1998). Once cured, the PDMS is peeled off revealing a stamp possessing the same patterns as the hard master (Figure 8.2) (Xia and Whitesides 1998).

8.2.1 Microcontact Printing

Two of the most common forms of soft lithography are microcontact printing and micromolding in capillaries (Lyman et al. 2010). Microcontact

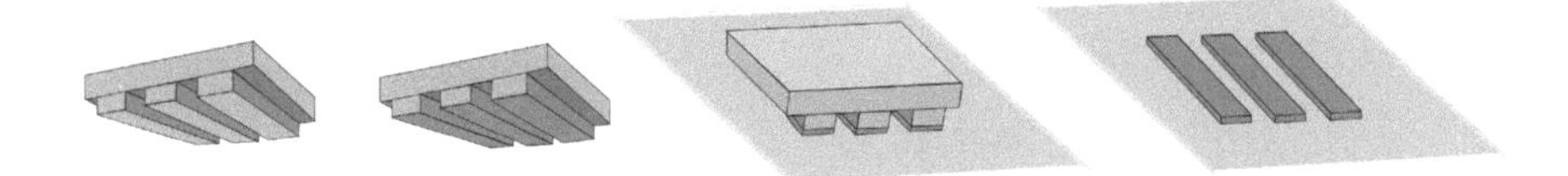

FIGURE 8.3
Microcontact printing procedure.

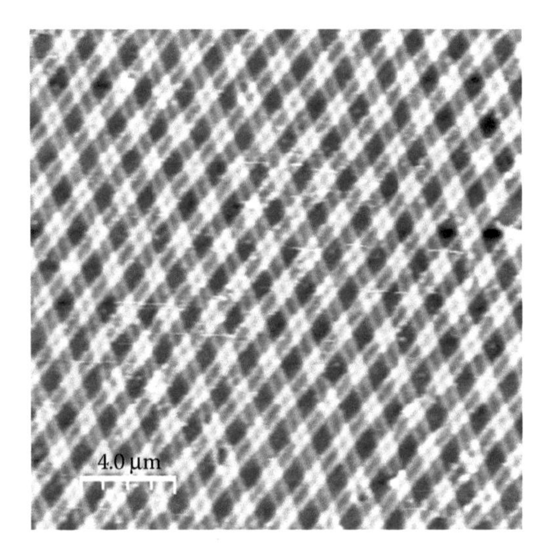

FIGURE 8.4
Polyvinylpyrrolidone grid on glass created using microcontact printing.

printing involves the transfer of molecular patterns onto the surface of substrates by making direct contact between an inked stamp and a flat substrate (Figure 8.3). The most commonly used "ink" for microcontact printing is an alcohol-based solution of alkanethiols (Lyman et al. 2010). This technique involves the use of protrusions on the surface of an inked elastomeric stamp to pattern structures on flat substrates (Figure 8.4) (Sanders 2015).

8.2.2 Micromolding in Capillaries (MIMIC)

Another common form of soft lithography is micromolding in capillaries (MIMIC) (Figure 8.5a,b). This technique involves placing a PDMS stamp with empty channels, pattern-side down, onto the surface of a flat substrate. A liquid is introduced to the open ends of the channel, and the liquid spontaneously fills the empty channels via capillary action. Next, the liquid is cured, or the solvent is evaporated, and the solidified material is confined in the channels. Once the stamp is removed, a pattern is left on the substrate surface (Figure 8.6) (Lyman et al. 2010).

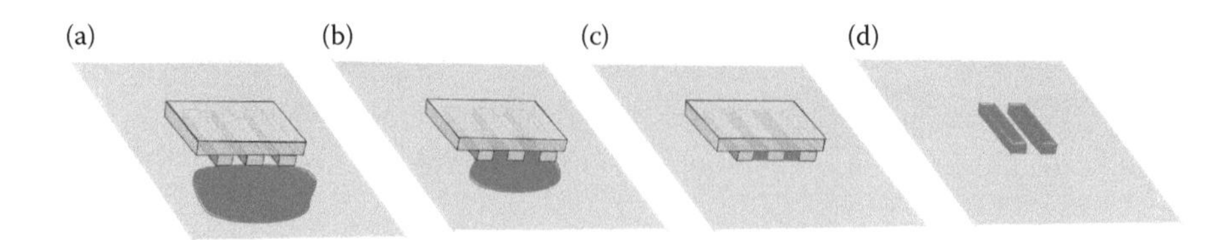

FIGURE 8.5
Micromolding in capillaries (MIMIC) procedure. A bare stamp (a) is coated with a molecular ink (b). The "inked" substrate is placed on a substrate (c) and once removed, the pattern remains (d).

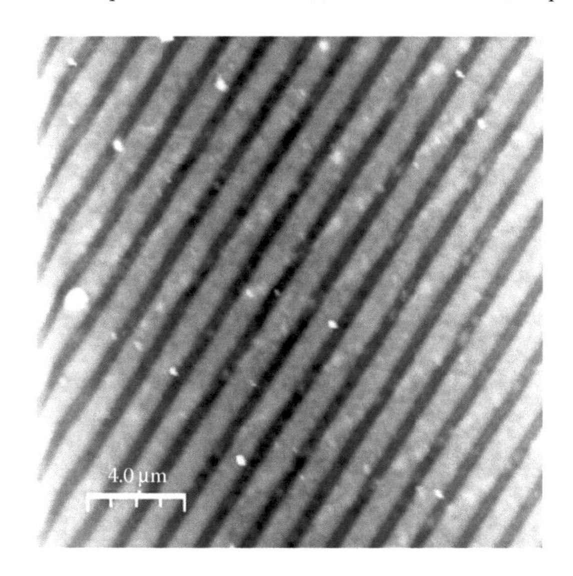

FIGURE 8.6
AFM image of polyethylenedioxythiophene (PEDOT) structures created using MIMIC.

8.3 Physical Vapor Deposition

Thin-film deposition is extensively used in micro/nanofabrication and is primarily used to deposit metals (Ziaie, Baldi and Atashbar 2010). In physical vapor deposition, the material is transported from a source to the wafers, both being in the same chamber (Ziaie, Baldi and Atashbar 2010). Physical vapor deposition (PVD) is a vaporization technique which involves atomic level transfer of material. This process occurs in a sequence of steps: (1) the material to be deposited is converted into a vapor; (2) the vapor is transported across a region of low pressure from source to substrate; and (3) the vapor undergoes condensation on the substrate forming a thin film. PVD processes can deposit films with thicknesses in the range of a few nanometers to thousands of nanometers (Liu, Ji and Shang 2010).

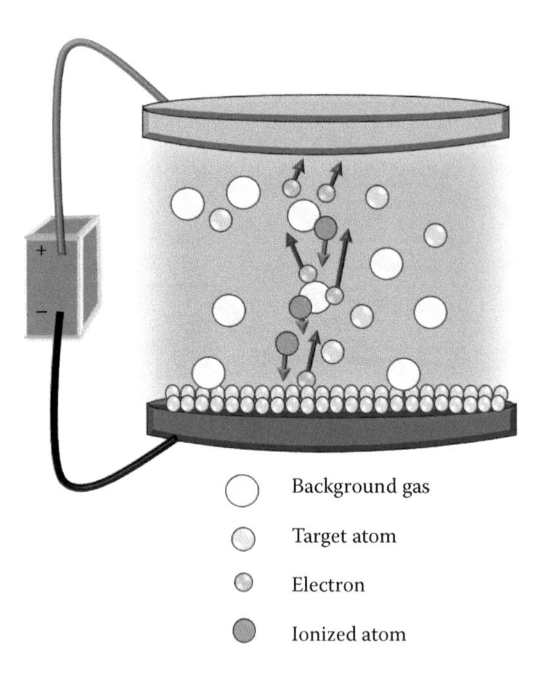

FIGURE 8.7
Sputter coating process.

8.3.1 Sputtering

In sputtering, a target material is bombarded with argon ions; this action removes atoms from the target and eject them towards a substrate (Figure 8.7). The argon ions striking the target are produced using a direct current (DC) or RF plasma (Ziaie, Baldi and Atashbar 2010). Sputtering allows better control of the composition of multicomponent films, and a greater flexibility in the types of materials deposited (Liu, Ji and Shang 2010). Plasma-based sputtering is the most common form of sputtering, using positive ions accelerated to the target, which is held at a negative potential with respect to the plasma (Liu, Ji and Shang 2010).

8.3.2 Metal Evaporation

In metal evaporation, the source material is placed in a small container (crucible) and is heated to a temperature facilitating evaporation (Figure 8.8) (Ziaie, Baldi and Atashbar 2010). Heated metals possess high vapor pressures, and in high vacuum, the evaporated atoms are transported to the substrate. The target material and the substrate are both located in a vacuum chamber (Liu, Ji and Shang 2010).

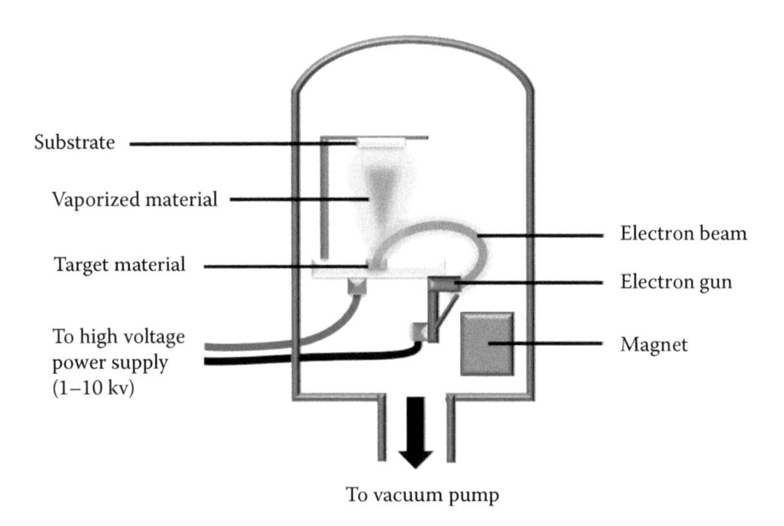

FIGURE 8.8
Metal evaporation system.

8.4 Chemical Vapor Deposition

Chemical vapor deposition (CVD) is a method of creating nanomaterials using a reaction involving vapor-phase compounds containing the required components (Liu, Ji and Shang 2010). CVD involves flowing a precursor gas or gases into a furnace containing one or more heated objects to be coated (Figure 8.9) (Creighton and Ho 2001). The gaseous source materials flow near a heated substrate; the source materials decompose and react

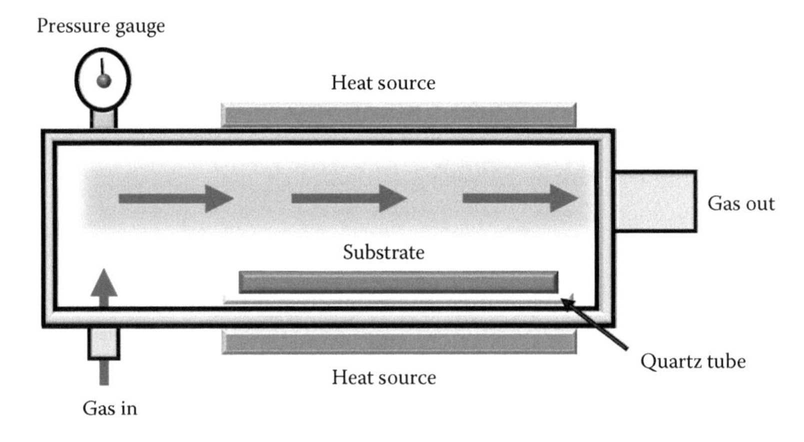

FIGURE 8.9
CVD furnace.

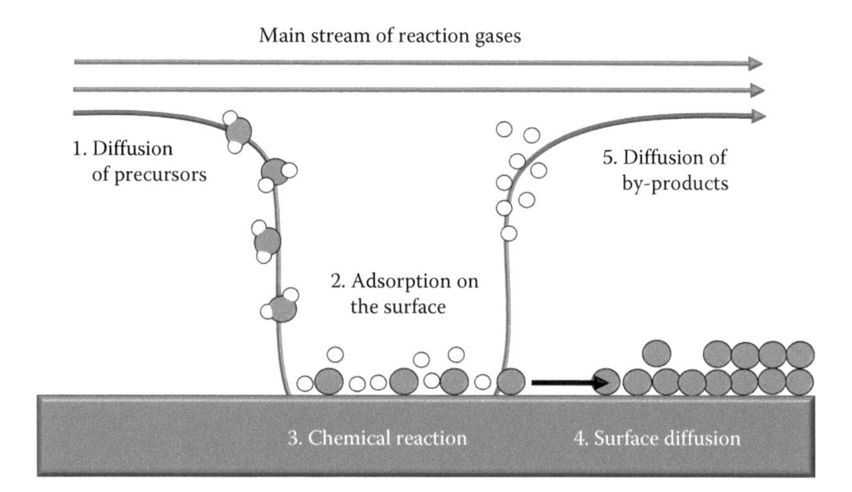

FIGURE 8.10
Mechanism for CVD.

leading to the fabrication of nanostructures (Liu, Ji and Shang 2010). This is accompanied by the generation of by-products, which are vented out of the furnace along with precursor gases that do not react (Creighton and Ho 2001; Liu, Ji and Shang 2010). The gaseous precursor is typically a hydrocarbon gas. This gas decomposes either thermally or in the presence of a plasma; decomposition is generally aided by a transition-metal catalyst (Bell et al. 2006). Additionally, CVD processes include low-pressure CVD (LPCVD) and plasma-enhanced (PECVD). LPCVD is usually performed in electrically heated tubes connected to pumps used to achieve the low pressures needed (0.1–1.0 Torr). The temperatures involved with LPCVD are generally in the 550°C–900°C range. PECVD uses radiofrequency (RF) energy to create highly reactive species in the plasma. This allows the use of lower temperatures (Ziaie, Baldi and Atashbar 2010). Energy in the plasma replaces some of the heat energy, allowing decomposition of the precursor gas to occur at lower temperatures (Bell et al. 2006). The CVD process occurs in a sequence of steps, as shown in Figure 8.10. Precursor gases are introduced into the furnace. The gaseous precursors are decomposed using heat, plasma, or other means. Atoms or molecules are adsorbed on the substrate surface and react to form nanostructures. Finally, byproducts are vented out of the furnace (Liu, Ji and Shang 2010). The decomposition of the precursor gas is aided by metal catalysts. The catalyst usually consists of a thin film of transition metals (Fe, Ni, Co, or Mo). The catalyst layer can be patterned using standard lithographic techniques resulting in highly localized growth of carbon nanotubes (Bell et al. 2006). Using CVD to produce carbon nanotubes involves the decomposition of acetylene gas over iron particles at 700°C. Substantial amounts of carbon nanotubes are produced by the decomposition of acetylene over iron catalysts deposited on silica (Wilson et al. 2002).

8.5 Etch

Etching is a process which creates topographical surface features by selectively removing material through physical or chemical means (Betancourt and Brannon-Peppas 2006). Etching mechanisms utilize liquid or gas-based processes (Betancourt and Brannon-Peppas 2006). Etching is isotropic if it occurs uniformly in all directions. Etching is anisotropic if it occurs in a specific direction (Betancourt and Brannon-Peppas 2006). Isotropic etches create a semicircular profile under the mask (Figure 8.11). The resulting features have curved walls and are wider than the opening of the mask (Liu, Ji and Shang 2010). In contrast, in an anisotropic etch, the dissolution occurs faster in the vertical direction, resulting in an etch with straight side-walls (Figure 8.12) (Ziaie, Baldi and Atashbar 2010). There are two major types of etching: wet etching and dry etching (Liu, Ji and Shang 2010). Wet etching involves immersing substrates in a reactive solution (etching) to remove materials from the surface of the wafer. Wet etching occurs in three steps: (1) diffusion of the etchant solution to the substrate surface; (2) reaction between the etchant and the substrate; (3) and diffusion of byproducts from the surface (Liu, Ji and Shang 2010). Undercutting, however, limits the minimum feature size achievable with wet etchants to 3 μm (Ziaie, Baldi and Atashbar 2010).

Dry etching techniques are plasma-based and possess several advantages when compared to wet etching. Dry etches are anisotropic and possess vertical sidewalls (Figure 8.12).

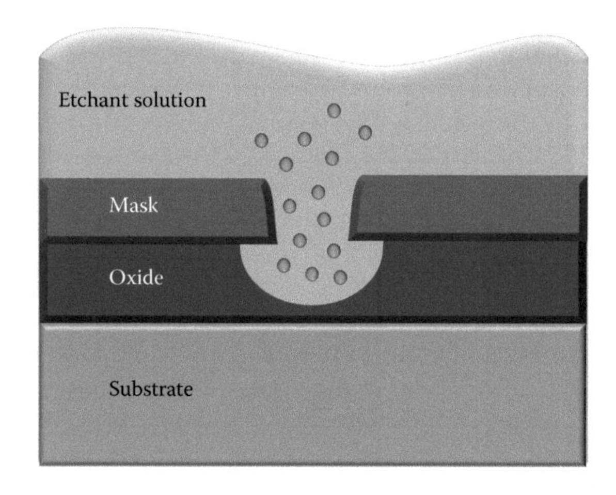

FIGURE 8.11
Isotropic etch created using wet etching.

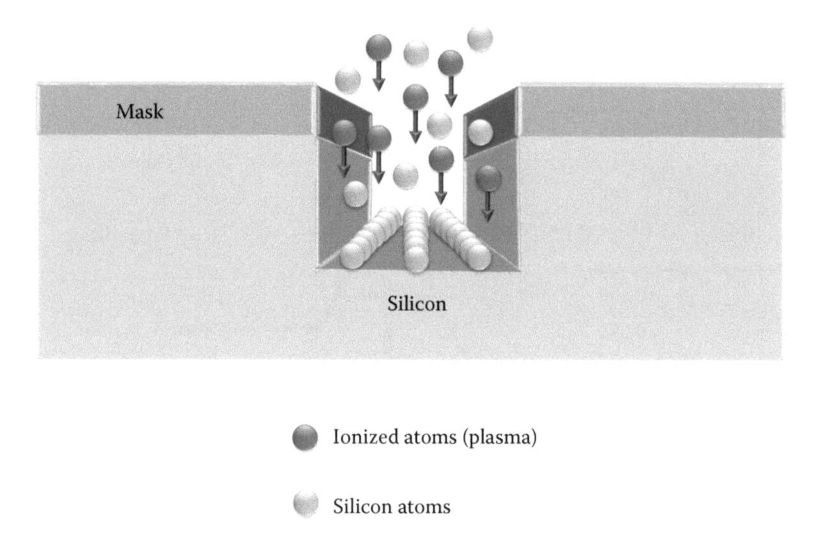

FIGURE 8.12
Anisotropic etch created using dry etching.

Dry etching offers smaller undercut, which allows smaller lines to be produced. Also, dry etching eliminates the need for dangerous acids and solvents. In addition, directional etching is possible with dry etching (Liu, Ji and Shang 2010). Dry etching occurs when incident ions from a plasma collide with wafers, breaking bonds at the surface. Ion collisions occur more frequently on surfaces than on the walls, resulting in faster etching rates in the vertical direction (Ziaie, Baldi and Atashbar 2010). Wet and dry etching involves the transfer of mask patterns onto a wafer. Masks ensure that only uncovered areas are etched away, while leaving areas covered by the masks intact (Liu, Ji and Shang 2010).

8.6 E-beam Lithography

E-beam lithography has been used to fabricate devices including integrated circuits, photonic crystals, and channels for nanofluidic devices (Altissimo 2010). Lines 5 to 7 nm wide have been patterned in a resist using an electron beam (Betancourt and Brannon-Peppas 2006). The working principle of e-beam lithography is simple. An electron beam is scanned across a substrate covered by an electron sensitive material known as a resist; the solubility of the resist changes when exposed to an electron beam (Altissimo 2010). E-beam lithography systems closely resemble SEM. However, electron beam lithography is scanned onto the sample in accordance with instructions from a computer-driven pattern generator. E-beam lithography systems consist of a chamber, an electron gun,

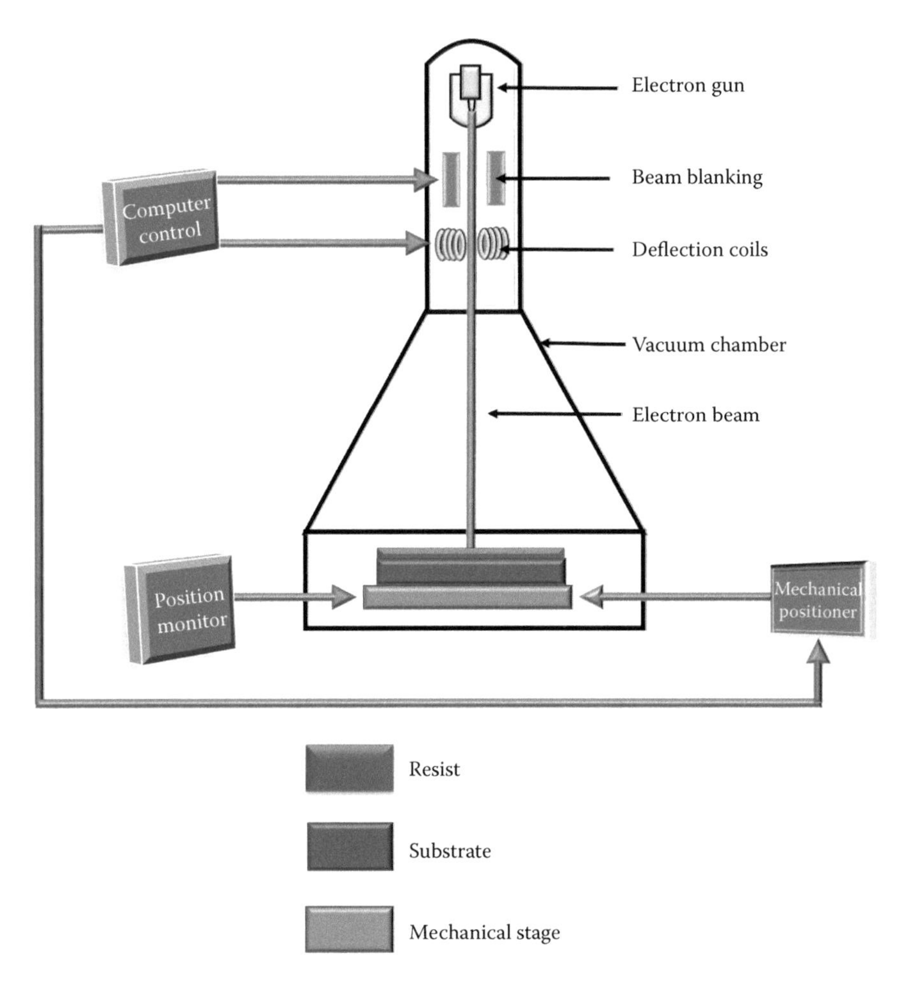

FIGURE 8.13
E-beam lithography system.

and a column (Figure 8.13). The column and chamber are maintained in high vacuum. Additionally, the column contains all the electromagnetic lenses used to focus the electron beam (Altissimo 2010). E-beam lithography is used in addition to processes such as lift-off, etching, and electrodeposition to fabricate various nanostructures (Altissimo 2010). The resolution obtained in e-beam lithography is influenced by the beam spot size. Various researchers have demonstrated resolutions on the order of less than 10 nm (Betancourt and Brannon-Peppas 2006). Polymethylmethacrylate (PMMA) is commonly used as a resist in e-beam lithography. The chemical bonds in a positive resist are weakened when exposed to an electron beam. Negative resists are cross-linked when exposed to an electron beam. Resist areas exposed—or not exposed—to an electron beam are removed by a developer solution (Figure 8.14) (Altissimo 2010).

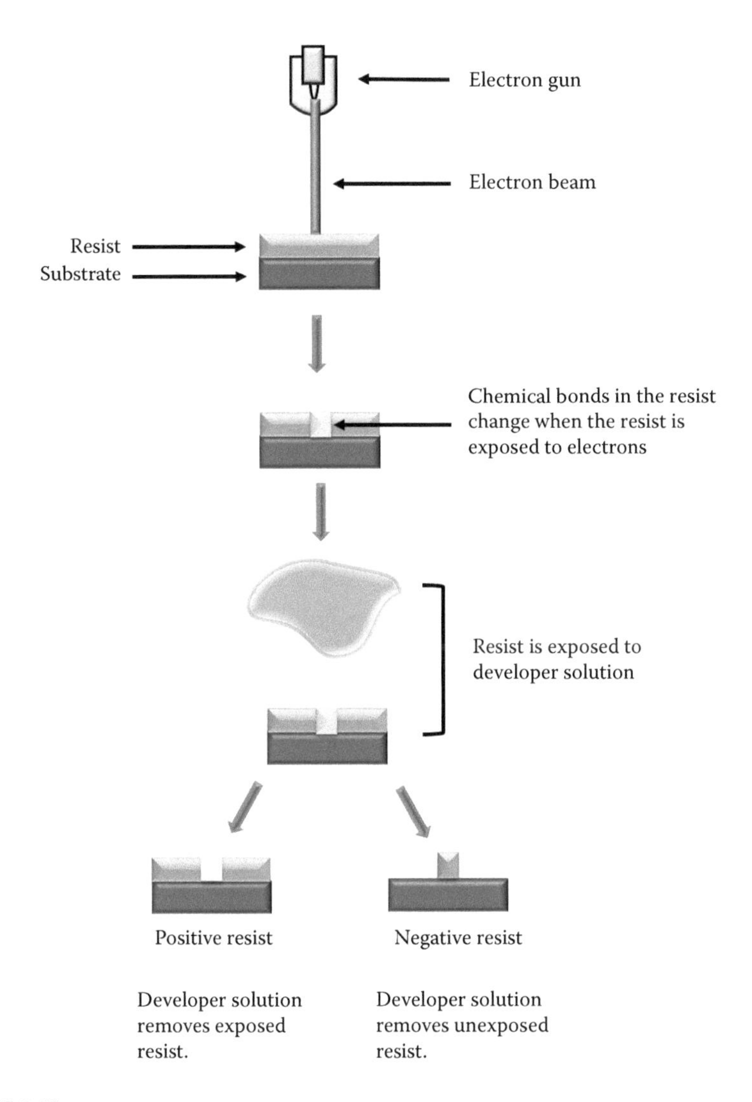

FIGURE 8.14
Behaviors of positive and negative resists when exposed to an electron beam.

8.7 Focused Ion Beam

Focused ion beam (FIB) has developed into a very attractive tool for lithography, etching, and deposition. FIB has also been applied for mask and circuit repair and can achieve a resolution as low as 100 nm. FIB components are shown in Figure 8.15. Heavy-ion species, such as gallium ions, are used in FIB (Liu, Ji and Shang 2010). Gallium ions are produced using a liquid metal ion source (Pradeep 2007). FIB is primarily used for the milling of samples,

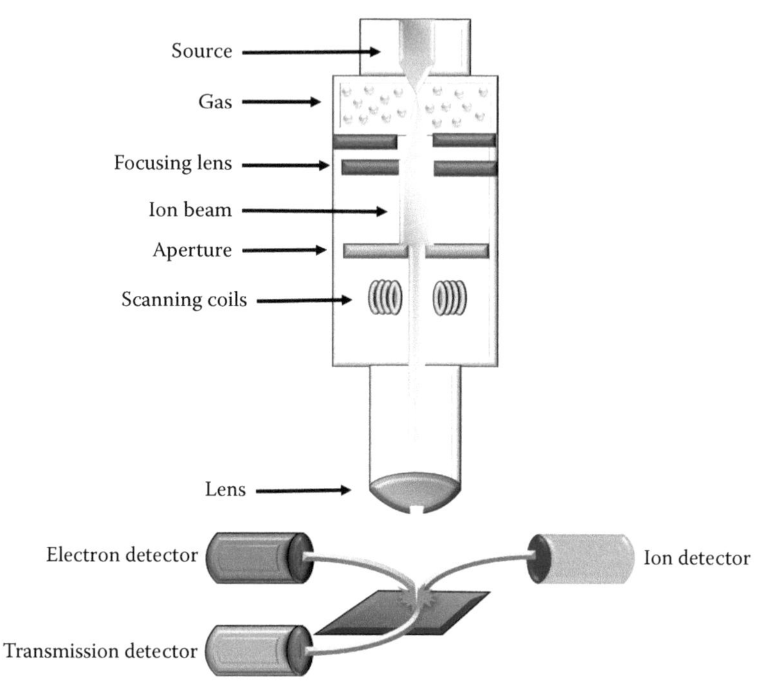

FIGURE 8.15
Focused ion beam (FIB) instrument.

however, the collisions between ions and sample surfaces can produce secondary electrons. This allows dual beam instruments to possess FIB and SEM characteristics, offering lithography and characterization capabilities in a single instrument (Pradeep 2007).

8.8 Photolithography

Photolithography is responsible for both the decrease in size and the increase in power of computing systems because it allows the incorporation of more components in an integrated circuit (Kaiser and Kuerz 2008). Photolithography is a process that is photographic in nature; it involves the projection of light onto a mask containing a pattern of an electronic circuit. Once light passes through the mask, the pattern is then projected onto a wafer covered with a light-sensitive photoresist (Figure 8.16) (Kaiser and Kuerz 2008). To obtain higher resolutions, higher energy (shorter wavelength) radiation is necessary (Wilson et al. 2002). The photolithographic process consists of many steps in which a desired pattern is generated on the surface of a substrate through exposure of

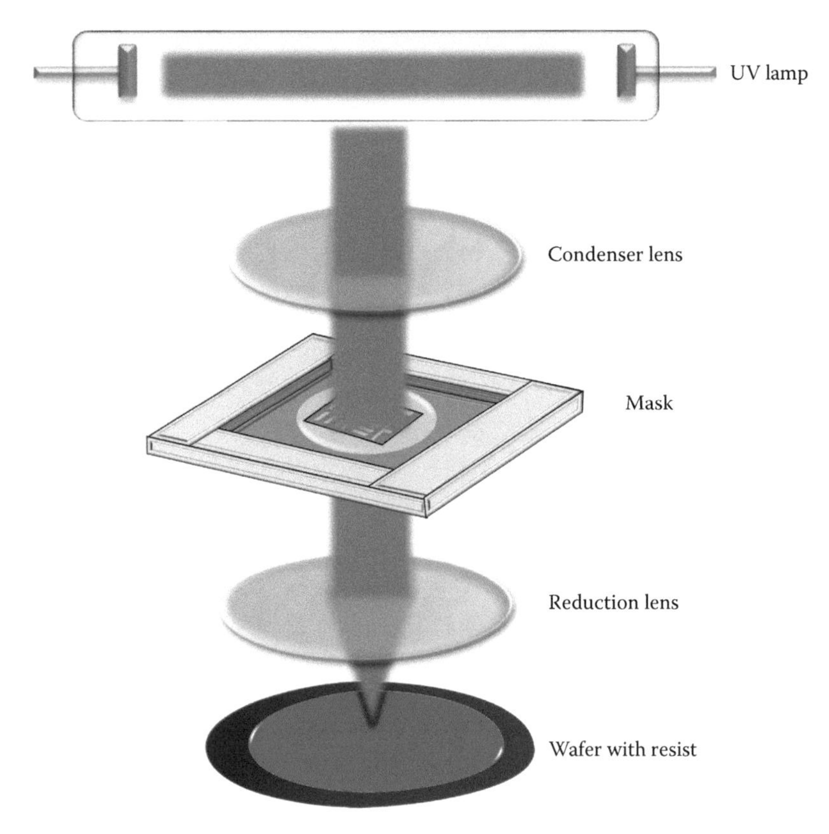

FIGURE 8.16
Photolithography assembly.

regions of a light-sensitive material to ultraviolet (UV) light (Betancourt and Brannon-Peppas 2006). The resulting photoresist patterns are then used to protect the covered substrate from etching, or from the deposition of compounds or biomolecules on its surface (Betancourt and Brannon-Peppas 2006).

8.8.1 Spin Coating

Spin coating involves coating a wafer by dispensing photoresist onto a wafer lying on a spinner (Figure 8.17). A vacuum chuck holds the wafer in place. Addition, a speed of about 500 rpm is used during the dispensing step, allowing the resist to spread over the substrate (Madou 2011).

8.8.2 Alignment and Exposure

A photomask is placed on top of the photoresist-coated wafer (Betancourt and Brannon-Peppas 2006). The photoresist-coated wafers, while inside an illumination or exposure system, are aligned with the features on the mask.

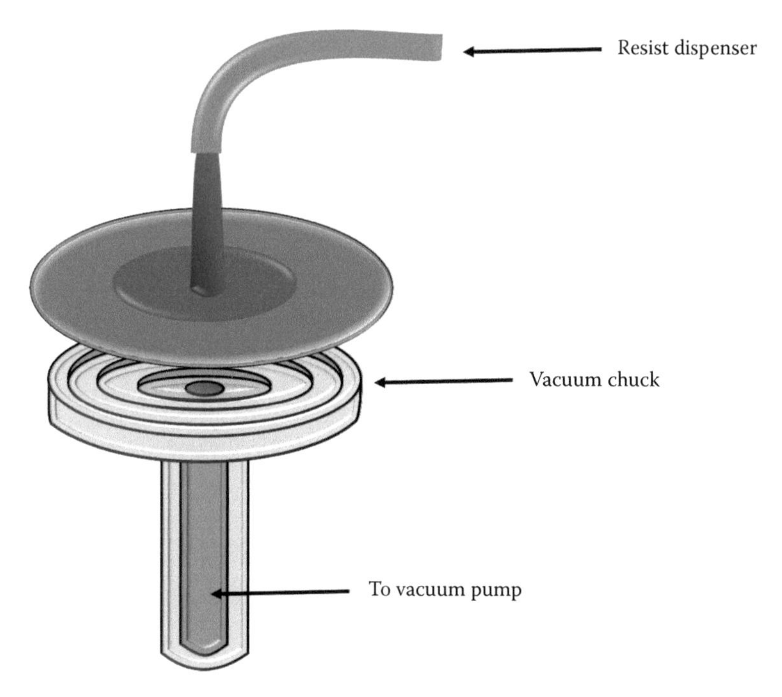

FIGURE 8.17
Spin coating resist on a wafer.

UV light with the proper intensity, directionality, spectral characteristics, and uniformity across the wafer facilitates transfer or printing of the mask pattern onto the resist. In photolithography, wavelengths of the light source used for exposure of the resist-coated wafer range from the very short wavelengths of extreme UV (10–14 nm) to deep UV (150–300 nm) to near UV (350–500 nm) (Madou 2011). When irradiated with UV light, sections of the photoresist not covered by the opaque regions of the photomask are exposed (Figure 8.18) (Betancourt and Brannon-Peppas 2006).

8.8.3 Development

Development is the dissolution of unpolymerized resist (Madou 2011). After exposure to UV light, the wafer is rinsed in a developing solution removing areas of photoresist and leaving behind a pattern of bare and photoresist-coated regions on the wafer (Madou 2011). Depending on the type of photoresist used, the photoresist will undergo one of two possible transformations when exposed to UV light. When light illuminates a positive photoresist, the exposed regions break down and become more soluble in a developing solution. As a result, the exposed photoresist can be removed when immersed in developing solution (Figure 8.19). A negative photoresist becomes cross-linked when exposed to UV light, becoming insoluble in the developing solution. Upon

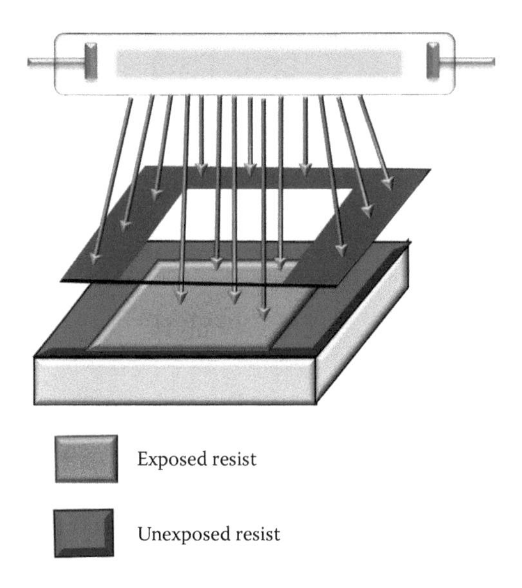

FIGURE 8.18
Exposure of resist to UV light.

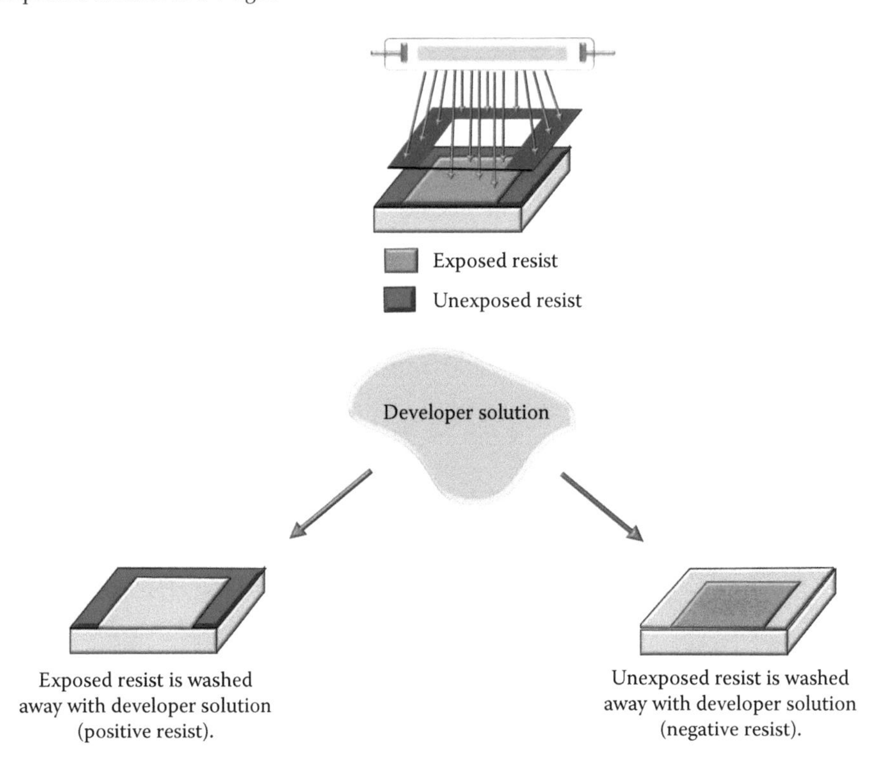

FIGURE 8.19
Reaction of exposed resist to developer solution.

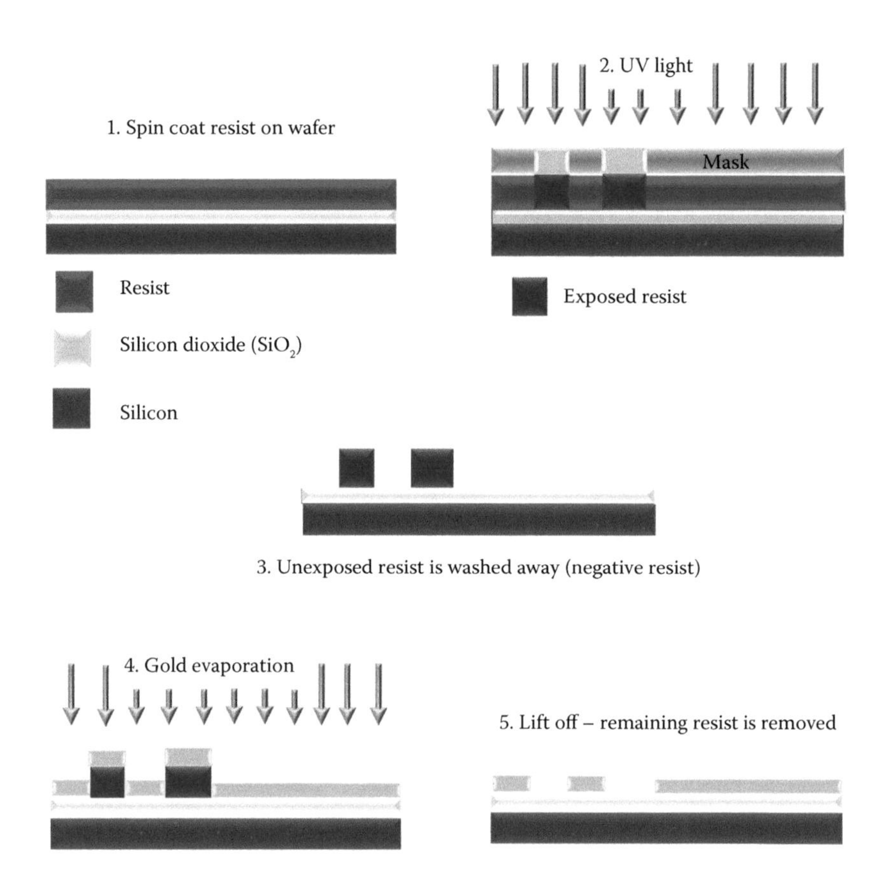

FIGURE 8.20
Fabrication of metal nanostructures using nanolithography.

contact with the developing solution, regions not exposed to light will be removed (Figure 8.19) (Betancourt and Brannon-Peppas 2006).

8.8.4 Lift-off

Lift-off is most commonly employed in patterning metal films for interconnections. The steps of the technique are shown schematically in Figure 8.20. After exposure to the developer solution, the wafer is coated with a thin layer of metal. Afterwards, an appropriate solvent (such as acetone) is used to remove the remaining parts of the photoresist and the deposited film atop these parts of the resist can be lifted off (Liu, Ji and Shang 2010).

8.8.5 Extreme UV Photolithography

To keep pace with the growing demands of manufacturing small feature size integrated circuits, extreme ultraviolet lithography was developed (Kaiser and Kuerz 2008). Extreme ultraviolet lithography differs considerably from conventional photolithography. There are no transparent materials for the small wavelengths used. For this reason, concave and convex mirrors are used. Each of the mirrors contains alternate layers of molybdenum and silicon layers, which will only reflect light with a wavelength of 13.5 nm (Kaiser and Kuerz 2008). In addition, extreme ultraviolet lithography is performed in a vacuum to prevent adsorption of the small wavelengths by gas molecules (Kaiser and Kuerz 2008). The central wavelength associated with extreme ultraviolet photolithography is 13.5 nm (Jonkers 2006). It is reported that feature sizes as small as 11 nm are possible with EUVL (Kaiser and Kuerz 2008). Laser-produced plasmas are used to generate wavelengths associated with EUVL. Laser-produced plasmas are generated when a target is irradiated with a pulsed laser source (Jonkers 2006). Electrical discharges in gases have also been used to generate wavelengths needed for EUVL (Jonkers 2006). Extreme ultraviolet lithography offers several advantages over conventional lithography. These advantages include:

- Production of minimum feature sizes
- Increased numbers of transistors in integrated circuits resulting in devices with greater speeds
- Creation of features with good resolution due to the small wavelength of UV light used (Chopra 2014).

8.9 End-of-Chapter Questions

1. The use of atoms or molecules to produce nanoscale materials is described as:

   ___ bottom-up nanofabrication

   ___ top-down nanofabrication

2. Soft lithography involves the use of a/an _____ to carry out micro- and nanofabrication.

   ___ electron beam

   ___ AFM tip

   ___ elastomeric stamp

   ___ STM tip

3. Elastomeric stamps used for micro- and nanofabrication are prepared using:

   ___ nanoshaving

   ___ a rigid master with a relief structure

___ dip pen nanolithography

___ nanoshaving and dip pen nanolithography

4. The form of soft lithography involving the transfer of patterns onto the surface of substrates by making direct contact between an inked stamp and a hard substrate is:

   ___ micromolding in capillaries (MIMIC)

   ___ microcontact printing

5. Solutions of alkanethiols are the most commonly used ink used in:

   ___ micromolding in capillaries (MIMIC)

   ___ microcontact printing

6. Filling the empty channels of a stamp on the surface of a hard substrate with a low-viscosity liquid for the purpose of micro- and nanofabrication is:

   ___ microcontact printing

   ___ micromolding in capillaries (MIMIC)

7. The nanofabrication method is a vaporization technique used to deposit metals using the transfer of material from a source to a substrate is:

   ___ physical vapor deposition (PVD)

   ___ chemical vapor deposition (CVD)

8. When a target material to be deposited is bombarded with high-energy ions, atoms are ejected and directed towards a wafer; this nanofabrication process is known as:

   ___ physical vapor deposition (PVD)

   ___ chemical vapor deposition (CVD)

9. What is the purpose of the ionized argon atoms in sputter coating?

   ___ dislodge metal atoms from a target

   ___ coat a substrate

   ___ break down a carbon containing gas

   ___ none of the above

10. Which of the following processes involves decomposition of a precursor gas for nanomaterial synthesis?

    ___ chemical vapor deposition (CVD)

    ___ physical vapor deposition (PVD)

11. Chemical vapor deposition (CVD) is a form of nanofabrication which involves the use of _____ phase chemicals.

    ___ solid

    ___ liquid

___ gas

___ solid and liquid

12. During the synthesis of carbon nanotubes via chemical vapor deposition, transition-metal catalytic nanoparticles are used to:

___ coat a substrate

___ dislodge metal atoms from a target

___ ionize argon gas

___ break down a carbon containing gas

13. Chemical vapor deposition (CVD) synthesis of carbon fibers, filaments, and whiskers involves the use of _____ and metal catalysts.

___ hydrocarbons

___ metal hydroxides

___ metal oxides

___ metal carbonates

14. Carbon nanotubes can be formed by catalytic decomposition of _____ using _____ catalysts.

___ carbon dioxide, copper (Cu) and zinc (Zn)

___ carbon dioxide, cobalt (Co) and iron (Fe)

___ acetylene, copper (Cu) and zinc (Zn)

___ acetylene, cobalt (Co) and iron (Fe)

15. Etches can be produced with reactive gases (plasmas).

___ true

___ false

16. Etching is _____ if it proceeds equally in all directions.

___ isotropic

___ anisotropic

17. Etching is _____ if it proceeds in one specific direction.

___ isotropic

___ anisotropic

18. Wet etches are _____.

___ anisotropic

___ isotropic

19. Dry etches are _____.

___ anisotropic

___ isotropic

20. Which etch technique can result in detrimental undercutting below the mask?
    ___ wet etch
    ___ reactive ion etch (dry etch)
21. The most common resist layer used in e-beam lithography is:
    ___ polydimethylsiloxane
    ___ polymethylmethacrylate
    ___ polyvinylpyrrolidone
    ___ polystyrene
22. In e-beam lithography, what changes the solubility of the resist?
    ___ ions
    ___ an electron beam
    ___ plasma
    ___ neutral argon atoms
23. Pattern generation in e-beam lithography is achieved using:
    ___ a mask
    ___ microcontact printing
    ___ micromolding in capillaries (MIMIC)
    ___ a computer
24. E-beam lithography can be used in conjunction with:
    ___ lift-off
    ___ etching
    ___ metal deposition
    ___ all of the above
25. Focused ion beam (FIB) uses _____ for lithography.
    ___ electrons
    ___ gallium (Ga) ions
    ___ plasma
    ___ metal atoms
26. Focused ion beam (FIB) can be used for direct milling and sputtering.
    ___ true
    ___ false
27. Dual beam focused ion beam (FIB) instruments can be used for:
    ___ lithography
    ___ imaging

___ characterization

___ all of the above

28. Which of the following industrial nanofabrication techniques involves a resist layer?

    ___ e-beam lithography

    ___ reactive ion etch (dry etch)

    ___ photolithography

    ___ all of the above

29. Regions of resist exposed to an electron beam are removed after exposure to developer solution. What type of resist was used?

    ___ positive

    ___ negative

30. Regions of a resist not exposed to an electron beam are removed after exposure to a developer solution. What type of resist was used?

    ___ positive

    ___ negative

31. In photolithography, pattern generation is accomplished by exposing a light-sensitive material to:

    ___ infrared radiation

    ___ red light

    ___ green light

    ___ ultraviolet light

32. What process is used to deposit a light-sensitive resist on a wafer during photolithography?

    ___ microcontact principle

    ___ micromolding in capillaries (MIMIC)

    ___ spin coating

    ___ sputtering

33. In photolithography, a developing solution is used to remove _____ resist when a negative resist is used, and is used to remove _____ resist when a positive resist is used.

    ___ exposed, unexposed

    ___ unexposed, exposed

34. What photolithographic process is used to remove the remaining resist after metallization or etch steps?

    ___ develop

    ___ alignment and exposure

___ lift off

___ spin coating

35. Although X-rays have shorter wavelengths that ultraviolet light, X-rays cannot be used in photolithography systems because:

___ of the lack of refractive X-ray optics

___ of the lack of reflective X-ray optics

___ of the lack of refractive and reflective X-ray optics

___ X-rays cannot induce chemical changes in resists

36. A _____ wavelength of light is needed to print more structures on a silicon wafer.

___ long

___ short

37. What is used to focus light in extreme ultraviolet lithography (EUVL) systems?

___ concave lenses

___ convex lenses

___ concave and convex lenses

___ mirrors

38. The smallest structure that can be printed using extreme ultraviolet lithography is:

___ 200 nm

___ 100 nm

___ 50 nm

___ 20 nm

39. UV light is produced in EUVL systems using:

___ lasers

___ xenon gas

___ plasmas

___ all of the above

40. What is the wavelength of light used in EUVL systems?

___ 13.5 nm

___ 75.5 nm

___ 150.5 nm

___ 293.5 nm

41. What elements are used to coat mirrors used in EUVL?

___ silicon

___ molybdenum

___ molybdenum and silicon

___ PMMA

42. To prevent the absorption of 13.5 nm ultraviolet light by gas molecules, extreme ultraviolet lithography occurs in:

___ nitrogen

___ a vacuum

___ helium

___ argon

References

Altissimo, M. 2010. "E-beam lithography for micro-/nanofabrication." *Biomicrofluidics* 4: 026503-1–026503-6.

Bell, M. S., K. B. K. Teo, R. G. Lacerda, W. I. Milne, D. B. Hash, and M. Meyyappan. 2006. "Carbon nanotubes by plasma-enhanced chemical vapor deposition." *Pure Appl. Chem.* 78: 1117–1125.

Betancourt, T., and L. Brannon-Peppas. 2006. "Micro- and nanofabrication methods in nanotechnological medical and pharmaceutical devices." *Int. J. Med. Sci.* 1: 483–495.

Chopra, J. 2014. "Analysis of lihtography based approaches in development of semiconductors." *Int. J. Comput. Sci. Inf. Technol. Adv. Res.* 6: 61–72.

Creighton, J. R., and P. Ho. 2001. "Introduction to chemical vapor deposition." In *Chemical Vapor Deposition*, edited by J. Park and T. S. Sudarshan, 1–22. Materials Park: ASM International.

Jonkers, J. 2006. "High power extreme ultra-violet (EUV) light sources for future lithography." *Plasma Sources Sci. Technol.* 15: S8–S16.

Kaiser, W., and P. Kuerz. 2008. "EUVL—Extreme Ultraviolet Lithography." *Optik & Photonik* 3: 35–39.

Liu, M., Z. Ji, and L. Shang. 2010. "Top-down fabrication of nanostructures." In *Nanotechnology: Volume 8: Nanostructured Surfaces*, edited by L. Chi, 1–47. Weinheim: Wiley-VCH.

Lyman, B. M., O. J. Farmer, R. D. Ramsey, S. T. Lindsey, S. Stout, A. Robison, H. J. Moore, and W. C. Sanders. 2010. "Atomic force microscopy analysis of nanocrystalline patterns fabricated using micromolding in capillaries." *J. Chem. Educ.* 89: 401–405.

Madou, M. J. 2011. *Manufacturing Techniques for Microfabrication and Nanotechnology.* Vol. 11. Boca Raton: CRC Press/Taylor and Francis.

Pradeep, T. 2007. *Nano: The Essentials Understanding Nanoscience and Nanotechnology.* New Delhi: Tata McGraw-Hill.

Sanders, W. C. 2015. "Fabrication of polyvinylpyrollidone micro-/nanostructures using microcontact printing." *J. Chem. Educ.* 92: 1908–1912.

Wilson, M., K. Kannangara, G. Smith, M. Simmons, and B. Raguse. 2002. *Nanotechnology - Basic Science and Emerging Applications.* Boca Raton: Chapman and Hall/CRC Press.

Xia, Y., and G. M. Whitesides. 1998. "Soft lithography." *Annu. Rev. Mater. Sci.* 28: 153–184.

Ziaie, B., A. Baldi, and M. Z. Atashbar. 2010. "Introduction to micro-/nanofabrication." In *Springer Handbook of Nanotechnology*, edited by B. Bhushan, 231–269. Heidelberg: Springer.

Index

Taylor & Francis eBooks

www.taylorfrancis.com

A single destination for eBooks from Taylor & Francis with increased functionality and an improved user experience to meet the needs of our customers.

90,000+ eBooks of award-winning academic content in Humanities, Social Science, Science, Technology, Engineering, and Medical written by a global network of editors and authors.

TAYLOR & FRANCIS EBOOKS OFFERS:

- A streamlined experience for our library customers
- A single point of discovery for all of our eBook content
- Improved search and discovery of content at both book and chapter level

REQUEST A FREE TRIAL

support@taylorfrancis.com